Glimpses of Gardens in Eastern China

著 | 童寯

译 | 童明

湖南美术出版社

情趣在此之重要,远甚技巧与方法。

——童寯

苏州……114　昆山……189
吴江……162　松江……190
扬州……164　青浦……192
泰州……174　嘉定……194
如皋……176　江阴……196
南京……178　太仓……198
上海……182　常熟……199
南翔……184　嘉兴……204
无锡……186　海盐……208
常州……188　平湖……210

Glimpses of Gardens in Eastern China ……217

附录 Appendix ……268

目录

序一　郭湖生　读童寯先生遗著『东南园墅』………… 1

序二　王澍　只有情趣 ………… 5

序三　董豫赣　造园与建筑 ………… 15

前言　中国东南园墅瞥观 ………… 21

一　如画之园林 ………… 24

二　园林与文人 ………… 34

三　建筑与布局 ………… 44

四　装修与家具 ………… 52

五　叠石 ………… 62

六　植物配置 ………… 74

七　东西方比较 ………… 90

八　沿革 ………… 104

(一) 历史 ………… 105

(二) 现况 ………… 112

◇ 童寯

（一九三五年摄于吴江）

序一 郭湖生

读童寯先生遗著"东南园墅"

童寯先生是近代研究中国古典园林的第一人。早在一九三一年，他决意遍访江南名园，独自手摹步测，数载惨淡经营，于一九三七年写成了《江南园林志》这部划时代的著作。而本书，英文版《东南园墅》，则是他于一九八三年临终前在病榻上亲手校订完稿的。童寯先生始终不渝致力于研究中国园林遗产达半世纪之久，他的贡献是不朽的。

用英文写作本书，目的是为了向世界介绍中国园林艺术。鉴于东西方文化体系之间的差异和隔阂，对于外国人来说（甚至许多中国人也一样），要欣赏中国园林，多少要了解生长它的深厚的中国传统文化土壤，要理解它所包含的中国哲学和美学内涵，要懂得它特有的风格和意境。情景交融，物我为一，没有情的会心，则于景也难以领略。寓情于景，触景生情，中国写景的诗、画，就是这一过程的凝聚物，中国的园林艺术也是这样，它们是同根所生，彼此之间易于触类旁通。

本书的写作，有一个很长的发展过程。最初，一九三七年顷，先生应《天下月刊》(T'ien Hsia Monthly)（主编林语堂、全增嘏）约，为该刊用英文写三篇文章，介绍中国文化，其中一篇为"Chinese Gardens"，但未发表。一九四四年，在重庆的一些学者拟合写一本英文专册"Chinese Culture Series"，一共八节。童先生分任其二："Chinese Painting"及"Chinese Garden Design"，均已完稿。然以他故，书终未成。但是"Chinese Garden Design"却成为本书的雏形。

一九七八年，刘敦桢先生所著《苏州古典园林》行将出版，并拟发国外版，先生为该书国外版写了英文序言，题名"Soochow Gardens"，一九七九年，先生拟写《中国园林植物配置》一文，对中国园林中所植各色植物花卉，按字母顺序编列拉丁文学名，用以介绍中国园林中的园艺品种。

以上述各英文著作为基础，先生于一九八一年写成《东南园墅略》，嗣后改名为"东南园墅"（Glimpses of Gardens in Eastern China），即本书。

一九八二年，先生病重住院以及转诊北京，均携书稿自随，时加订正。一九八三年三月，先生于病榻上口授此书结尾部分，全书竟。两周后，先生溘然长逝。呜呼，先生可谓"鞠躬尽瘁、死而后已"矣。

为了对东方（包括日本）和西方的园林艺术做比较研究，先生多年来广泛阅读和积累世界范围的园林资料，其中包括他自己旅欧期间的亲身见闻体验。这方面的积累，在先生晚年，由于力不从心，没有能发展到应有的规模，而简直就以素材的形式发表了其中的一部分，如《造园史纲》。就如我所知，先生对于日本造园学十分重视，对于日本造园史也有深切的了解和独到的见解。对日本历史上的重要造园家及其事迹，历历如数家珍。先生是打算把中国园林艺术置于世界总体之中来观察和评价。很遗憾，为这个复杂的课题所做的一切努力，因他的去世而化为乌有。

当前，研究园林艺术在中国已经成为热门、显学，号称专家者比比。但是，依愚所见，具备有如童寯先生的渊博知识，足以胜任这样复杂课题的，目前似无第二人。童寯先生一生不声不响地工作，从不喜欢和记者打交道。我们应该学习他，多下功夫，少点喧嚣吧！

先生写作本书，有赖南京工学院（一九八八年更名为东南大学）晏隆余同志的悉心协助，举凡照相、绘制图表、补充资料及至身后出版事务，均由晏君任责操劳。他的贡献，应予表彰。

值遗著问世之际，书此以纪念童师。师孜孜终生，可以含笑瞑目矣。因述书之始末，抒所感所怀，以为记。[1]

郭湖生
一九八六年八月五日于灯下

---

1. 编者注：此文为《东南园墅》第一个中译本的序言。

序二　王澍

只有情趣

近日，好友童明将其祖父童寯[1]先生英文旧著《东南园墅》以中文重译，托我作序，我在讶异之余，欣然应允。

所以讶异，是因为《东南园墅》在二十世纪九十年代已有中文旧译，我应该是最早一批读到这个译本的几个读者之一。正式出版之前，当时负责编辑的东南大学出版社的晏隆余先生还曾就书中一些关于园林植物的照片内容托我帮助核对。我对江南旧园林的态度，从原来觉得老套重复且已经在今天失去意义到重新发生热情与兴趣，如果说阅读更早一些出版的《童寯文选》算是预热，

---

1. 编者注：此文中童先生皆指童寯。

那么读到一九九七年《东南园墅》第一个中文译本，就是真正的思想转折。有意思的是，《童寯文选》原文也是英文。一九八七年，在《童寯文选》出版之前，也是晏老师托我将译稿带去北京，让清华大学建筑系的汪坦先生帮助校对。我还记得汪先生家住在清华大学那栋很新潮的低层高密度阶梯住宅楼里，印象最深的是他家客厅里的一棵大树，已经撑满客厅，满墙书架的书房就被挤在向客厅敞开的一个三米见方的朝阳小间里。当时只是感觉特别好，现在回想起来，汪先生家的那一小方天地，何尝不是一处最小的文人园林呢？不过，那一天的印象不只是视觉的，汪先生已经看过我的硕士论文，是我去之前寄到他家里的。他很高兴，留我吃中饭，那是我第一次吃到西洋式烤面包，也是第一次见到烤面包机，现在的回忆里，嘴里似乎还带着汪先生家的面包香味。

  回忆这些并不是跑题，因为园林不只是视觉对象，更是身体经验，曾经的园林主人在园中经年度日，岁月悠悠，生活是有味道的。童寯先生的文章能让我重新发生对园林的兴趣，就在于他不是掉书袋，不是去解释，而是以一个出色建筑师的眼睛和身体去发现园林的意趣，这和建筑史研究的角度很不同。以往国内的园林研究，尽管最早是由童寯先生开启，但后来主要是建筑史学者在做，以记录、测绘和历史考证以及图像解释为主，这种研究里没有问题，只有所谓方法，意义逐渐封闭，甚至让人疲倦。我至今仍然记得，二十世纪八十年代末，当我在还未正式出版的《童寯文选》中读到"中国的园林建筑布置如此错落有致，即使没有花草树木，也成园林"这句话，心中那种豁然贯通的感觉。印象太深，写这句话时我没有查资料，是背诵的。这句话对做设计的建筑师是能够产生重大影响的，因为它带出了园林的抽象结构，使得园林语言和西方现代建筑语言之间形成可能的对话关系。让我印象深刻的另外一句，出自《东南园墅》，是一个问句，质问假山石上的洞，大意是"一个正常的人怎么能住进那么小的洞中？"。一九九七年，我第一次读到这句话，当时浑身一激灵，脑袋轰的一下。这个看似幼稚的问题，切中园林语言的特殊逻辑，这是我以往没有想到的。如果按解释的路子，可以说那么小的洞是给想象中的仙人住的，于是一定有人去翻阅古籍，去研究园林和道家仙人思想的关系，但童寯先生的这个问题了不得，它让我一下子理解到园林语言中那种小与大并存的矛盾的尺度逻辑。这种问题是建筑史学者不会问也问不出来的，只有像童先生那种对设计过程有深刻体会的建筑师才会问得出来。这种发问的方式对我的影响不仅止于此。实际上，我经常比较稚拙，譬如，我读童先生这句话时，脑海里就会出现丰子恺先生漫画般天真的场景：童先生站在园林里一座假山前，用眼睛望着，也用手指着那上面的一个异形小洞。问题是，这就点出了园林语言中视线和差异空间的现场关系，明白了这个，深浅、进退、开阖、高下、疏密、小大……这些和园林有关的术语才会有活的意义。有国外建筑学者曾经评价说我发明了一种特别和视线移动有关的建筑语言，应该说，这种思考，其实特别和中国山水画的绘画语言有关，而我对山水画产生新认识，最早就是被童寯先生的《东南园墅》一书内的问题给刺激出来的。童寯先生也是在这本书里，直接讨论了园林与山水画的关系。我没有考证过童先生的这个讨论是不是近代国内最早的，这对我也不重要，对我重要的是，童先生的讨论才是真正懂中国画的人的讨论，只有这种讨论才对我有意义。刘敦桢先生在《江南园林志》序中说童先生知六法，所言不虚。

另外一个有意思的问题是关于这种思考与发问的关系。无论《童寯文选》还是《东南园墅》，原文都是英文，前者写于二十世纪三十年代，童寯先生那时应该是在上海华盖建筑师事务所工作，利用周末去苏州访园，同行的经常有外国友人。可以想象，一定发生过非常有趣的讨论，刚才谈到的那个幼稚问题就相当可能和这种漫游有关，童先生这一批用英文写的讨论苏州园林的文章就是那个时候为上海的英文杂志写的。《东南园墅》写于新中国成立以后，应该是写得断断续续，最后完成要到二十世纪七十年代末。同样用英文写，既和二十世纪三十年代的那一组文章有继承关系，应该也和"文革"末期的处境有关。让我感慨的是，童先生这种天真发问的精神，居然能够从二十世纪三十年代一直保持到生命的最后。这就见出"情趣"一词的重要，一九九七年，我在把《东南园墅》反复读了六遍之后，兴致盎然，就把童先生的《江南园林志》拿出来重读，于是，"情趣"二字跃入眼帘，直中我心。我意识到，园林营造不从理论开始，不由方法左右，和重要与否无关，最重要的就是这两个字：情趣。建筑师的道路总是困苦艰难，什么能支撑你一直有感觉地做下去？是什么理论吗？重大价值吗？方法吗？我体会都不是，情趣，童先生说出的这两个字，轻轻飘飘，但最能持久，因为它活色生香，是不断生发的。

  我和童明成为好友，首先和他是一个好人有关，我们是同济大学一九九五年建筑博士班的同学，住同一栋宿舍。刚认识他时，他并不是一个有趣的人，也不喜欢建筑，典型"理工男"一个。如果说童先生的"情趣说"传染了我，那么我对重读童先生园林著作的热情也一定传染了童明。他后来就开始做整理出版童先生旧著的工作，也曾托我整理童先生《中国建筑史》和《中国雕塑史》两部书稿。我也很想做，但我实在是太懒散，不是这块料，很辜负朋友所托。而童明则逐渐对建筑发生了热情，对园林研究逐渐痴迷，并把童寯先生几乎所有旧著整理出版了一遍。难以想象这是怎样海量的工作，不佩服都不行。

《东南园墅》的第一个中文译本对我建筑思想的形成影响很大，童寯先生的英文原文也写得非常简洁清楚。第一个中译本的翻译可以说是忠实于原文的感觉的，所以我看到童明的新译才觉得讶异：译文有点像旧骈体文，对偶句、排比句连串，很多人也一样可能读不习惯。但我并没有急于判断，因为童明做学问一向是很严肃的。

我想到的第一个关键词是"翻译"，不是在一般意思上谈，而是从多个角度谈。童寯先生的原文都是英文，最早是写给外国人看的，中国人默认的文化语境在外国读者那里不管用了，怎么让读者理解？什么是跨文化、跨世纪的语境的解释？我前面说过，解释的方法越来越不管用，所以童寯先生用了一种发现式的讨论方法，一切就像是第一次看见一样。传统文化背景和建筑学专业背景当然也在起作用，但童寯先生的写作有一点我很认同，他在写作时非常有理论自觉，没有什么概念是不经讨论就拿来用的。如果说园林是立体的中国画，是时间加空间的四维诗歌，从这种意义上说，脱开原有文化语境的发现式写作，就像是一种创造性翻译，这是翻译诗歌时的办法。我对园林和中国画的一系列新看法就特别受益于童寯先生的这种"翻译"。

第二个关键词是"文体"。二十世纪人文社会科学的一个重大变化就是现代语言学观念引发的思想变革，法国结构主义和后结构主义哲学的一个深刻发现就是：在表达解释与意义之前，文体从结构的深刻层次预先决定着意义的语境，也就决定性地决定了意义。实际上，用什么文体讨论园林更合适，很少有人问，但它确实是一个严肃的学术问题。譬如，计成就认为用骈体文写《园冶》是最合适的选择，我猜测童明用这种接近骈体文的文体再次翻译，就有这种意图。但童明的文体并不是骈体文，结构像，但词汇简明，完全不是骈体文惯用的绮丽文字。反过来想，童明的新译肯定让我们意识到，童寯先生的英文文体本身就是对园林的一种文体介入，不管是自觉的还是不自觉的。

　　第三个关键词是"接受"。无论原文还是翻译，作者都不可能忽略读者的存在。童寯先生的英文原文最早就是写给外国人看的，就需要想办法让外国人读懂，反而就写得特别清晰明了。冯友兰著、涂又光译的《中国哲学简史》也是同一种情况，效果也不错。这种文体还有特殊的一点，好像总有人在不停发问，先生就耐心地回答，文字之间，即使没有用问答体，也可以体会到那种问答的语气。我不认为用英文或者第一个中译本翻译用的白话文写园林就不能达到园林的真意，实际上，童先生的英文就很有明朝小品文的气质，译文也把这层意思相当好地传达了出来。从这层意思上，童先生的写作在试图让人理解园林时，已经催生出了一些新东西。罗兰·巴特就认为，这种以理解性为目的的写作活动，其结果往往是新东西的诞生。就像童先生的文字在我身上发生的化学反应。巴特的"写作"概念特别强调这个，要求读者如写作一样阅读，这就对读者提出了更高的要求。当童先生的视线指向太湖石上的小洞时，那个看似不在的仙人实际上也看着他。不要忘记，读者作为旁观者也总是在场的。所以，我觉得童明新译的特殊文体意味着这不是一般的翻译，而是关于园林理解的一个新的文体实验,因为对今日的园林来说,骈体文和白话文一样遥远。

第四个关键词是"质地"。我不想在这里直接讨论童明新译的种种细节,避免在"接受"的意义上干扰读者的阅读乐趣,但我忍不住要指出他的新译本开始几句中一个词的特殊意味,那个词就是"弱径"。英文原文里是小径的意思,但童明肯定认为翻译成"小径"完全不能传达园林里那种路径质地的意思,这种质地不仅是指材料质感,我觉得,也指线条的笔感、状态,和文人画上那种虚实有致的意思对应,我们会说这是一种味道,一种情趣。童明选择这个词是自觉的,你可以同意也可以不同意,但这个翻译中的意义遗漏被他敏锐地指出了,第一个中译本译文的用词质地显然还不够,所以,可以想象童明在翻译过程中是如何字斟句酌、一个字也不放过的样子,这个译本见证了童明对园林的理解达到了一个新境界。也可以对比童寯先生在《江南园林志》里文采斐然的文字,去想象如果童寯先生自己用中文重新写一遍《东南园墅》,又会是什么结果。当"情趣"这个羸弱的词成为中心词汇,就可以理解童寯先生说园林不可度量是什么意思。不可度量就不可设计,的确,一条小径可以度量,一条"弱径"又如何度量呢?

当然，假山也难以度量，所以童寯先生也同意历来文人的见解，把堆假山列为园林中第一有难度的操作。但假山毕竟仍然可以操作，另外一种园林中真正难以度量的事物就是植物，我觉得，和上世纪三十年代童先生"中国园林没有植物也成园林"的判断相比，晚年的童先生对植物在园林中的地位的看法已经变化了，所以他才试图在《东南园墅》中补充关于植物的部分，尽管在前一个中译本中，那种罗列植物图片的方式未必是童先生的本意。而我对这个问题的理解则是，童先生最早在《江南园林志》一书的一张图片的图注文字里已经透露出一层特殊的意思，那张图片上起伏的曲墙接续了茂盛的紫藤，童先生写道："这墙在哪里结束？植物又从哪里开始？"如果说，植物和墙在园林的完整理解上是处在不可分的状态，那就意味着园林总是鲜活地生发着，这是一种难以被固定理解的特殊的建筑学。

按这个意思去读童明这个新译本，读者就一定惊喜连连。

王澍
二〇一八年四月十六日

序三 董豫赣

造园与建筑

一

每引童寯之景观草坪只能吸引奶牛之言，痛快之余，不免心虚，明明记得是夫子所言，却总想不起出处；每闻王欣以童寯之无花木亦可成园而起论，疑惑之际，搜肠刮肚，对童寯在哪里讲过这些话，竟全无印象。

后来私询童明才知，我与王欣对童寯各执一端的摘录，皆在《东南园墅》和《童寯文选》两本书里出现过。

我所一再引用的那段原文是"中国园林必不见有边界分明、修剪齐整之草坪，因其仅对奶牛颇具诱惑，实难打动人类心智"；王欣所摘录的那段原文是"中国园林中，建筑如此赏心悦目，鲜活成趣，令人轻松愉悦，即便无有花卉树木，依然成为园林"。

在这两本著作里，童寯皆以西方园林两条特征短语为引，展开相关庭园植物的中西比照：

针对法国一位诗人之言"吾甚爱野趣横生之园"，童寯力证中国庭园早已摆脱了山野丛林的荒蛮气息。当年读到童寯从凡尔赛宫修剪整齐的几何植物里窥见它从未消除的荒漠气息时，我一直困惑难解，按黑格尔的诠释，将自然进行几何景观的秩序化，正是为消除自然令人不安的荒蛮感，进而获得不被打扰的旁观心安。直到近年，我用如画旁观与入画居游的对照视野，重审中西方园林的使用区别时，才隐约理解其荒漠气息，大概是指其无关身体居游的旁观疏离吧。大概是在这个园林宜居的中国语境下，童寯才会得出近乎调侃的结论，西方的景观草坪只能吸引奶牛。

针对约翰逊博士之问"难道园林不都是植物园吗？"，童寯力证西方园林具有以植物种植为特征的丛林气质，无论是法国几何园林，还是英国如画园林，建筑与植物丛林的关系疏离（童寯以汪洋中的孤岛来比喻），景观要素之间的关系，远比景观与建筑之间的关系更为紧密，这与中国园林中建筑与林木之间构造出即景的紧密关系差异巨大;或是为对照出西方园林的植物园气质,童寯才会说,中国的园林虽无林木亦可成园。

二

在童寯的《造园史纲》里,也有一个"东西互映"的标题,但写法与《东南园墅》或《童寯文选》都很不同。他几乎是以同时性线索,不断比照中西方庭园的兴衰——古罗马与西汉规模宏伟的帝王苑囿与私家园墅几乎同时出现;当西方中世纪庭园没落成修道院庭园的一角草坪时,正是两晋到唐宋间中国山水庭园的成型期;在意大利文艺复兴的庄园重兴时,正值中国江南园林的勃发期;而在十七世纪前后涌现的造园家里,在日本有小堀远州,在中国有计成,在法国有勒诺特尔,在英国有布朗。

语及近代,情绪则从兴奋而低落,童寯一面对美国人发明的景观建筑学这个合成词表示乐观,认为造园与建筑,在艺术上息息相关,他给出的一张巴黎的现代悬空园案例,似乎正是柯布西耶的屋顶花园;另一方面,童寯一再将景观建筑学这个新兴学科,视为中国园林将要式微的潜在威胁。

或许，童寯创"园墅"一词，竟会意在用中国园林抵抗景观建筑学？

在《造园史纲》里，为比照两宋兴起的"城市山林"，童寯曾信手摘来一个拉丁同义词 Rus in Urbe 相映照，这条如今须一篇博士论文论证的发现，竟被童寯放在尾注里。但它见证了古罗马时期的城市生活曾有过与苏州庭园一样的自然场景。庞贝引入自然场景的城市庭园，并非由建筑专家或景观专家所设计，只是人们对人工与自然共栖的自然选择，但它随后被中世纪禁欲的火山灰掩埋千年。文艺复兴虽复兴了古希腊古罗马的建筑，却因建筑学的专业兴起，终将庭园内的景观草坪，视为建筑的配景，此后从这配景分裂而成的景观专业，又因各自独立，竟失去最初的基本观照。美国一百年前发明的景观建筑学专业，不但没能解决建筑与景观不能共栖的历史习惯问题，而且只能在相关生态与环保技术上，和景观与建筑专业争夺地盘。

正是在回溯古罗马庭园所处的不分专业的年代时，童寯第一次用中文写出"园墅"一词，并将它用于最后之作《东南园墅》的书名，或许他对与古罗马庭园更接近的中国园林抱着能抵抗景观建筑学侵蚀的期望，或许还有对中国建筑未来能园筑合一的期望。

"园筑"一词，是计成对庭园建筑的特指。我曾将日本由庭园建筑缘侧省略庭景而得的灰空间理论，视为建筑专业无力造景的被迫之举，继而将现当代建筑师普遍无能造庭院的情形，追溯到世界园林史的分工习惯——日本园林史上，建筑与景观，专业分工明确，它与西方园林史的近现代分工高度接近，却与计成要杂识堪舆、建筑、陈设、造景、绘画乃至文学的中国造园文化，相去甚远。

三

最后，《东南园墅》这本著作，是应保留更适合西方学者阅读的格式，以便向世界推广中国园林，还是将版式与文字都修改为适合中国学者阅读的风格，以对中国现代建筑与园林有所启示？这两者，对当代中国，都一样迫切。基于建筑师的身份，我暗自倾向后者，但新的阅读对象是园林专业，还是建筑专业？

童寯说"造园与建筑，在艺术上息息相关"，中国造园，很难区分为建筑与园林两个专业吧。在《东南园墅》里，就有这么一段文字：

"唯文人，而非园艺学家或景观建筑师，才能因势利导，筹谋一座中国古典园林。即便一名业余爱好者，虽无盛名，若具勉可堪用之情趣，亦可完成这一诗性浪漫之使命。"

对童寯的造园文人情趣说，王澍或许会心，而对童寯用业余爱好者来描述造园文人，自诩为业余建筑师的王澍，或许也受此影响；葛明语近偏颇的公开断言"不懂掇石者，不配讨论中国建筑"，大概会得到童寯这本书的宽容，与明人论园不同，童寯用叠石而非林木来调剂庭园内的人工与自然，大概是叠石与种植，还有是隶属建造还是园艺的细微分别。我这些年从留园鹤所一带让人迷恋的庭园场景内，意识到中国庭园以墙垣空窗杂景的操作潜力。它本是建筑师可堪营园处，而这一世界罕见的庭园墙垣之能，在童寯这本著作里，也得到了特殊观照。而我头些年见到王欣尚在北京建筑大学时所带的学生作业，最吸引我的不是建筑，而是建筑与空庭间杂间翠色的曲折墙垣，尽管我不清楚空庭中的翠绿，是草坪还是林木。

董豫赣

二〇一八年八月二十日

前言

中国东南园墅瞥观

    论及中国东南园墅，苏州园林实为中国传统景观艺术之最，已成普遍共识。就其一般且共同特征而言，一座苏州园林与该地区其他园林别无二致，由此，观赏一座苏州园林，犹如领略所有同类园林。苏州园林之称最天下，其因在于历史背景、高雅品味、众多数量。远溯至公元四世纪，苏州由于拥有顾辟疆园，名闻遐迩。然而自十一世纪以往，确切园址似已湮没于历史烟云而无从辨识。是园概为江南第一座有名之私家园林。

    现今所存最早之苏州园林，起始日期可溯至十世纪。园林历史愈久，因其不断更迭，与原设计之景象相去愈远。今日多数苏州园林，清代始建或重修，大体都在上世纪后半叶。苏州亦为地区之工艺中心，拥有砖工、木作等精湛技艺，另加运河与道路之便利交通，农业生产与商品贸易促成繁华之经济，并为文化活动奠定基础。宜人之气候，丰富之水源，共同促进园艺事业之发展。物华天宝之地，人杰地灵之所，有钱及有闲阶层趋而往之，群英荟萃，长少咸集。园匠、诗者与画师，无不竭尽才智、苦心经营，相地构园、营厅造轩、种梅植竹，以为怡情享乐之用，进而促成繁荣兴旺之园林。基于相同原因，其他城市亦有园林发展，却在数量方面与苏州不可相比。

扬州园林于数量方面仅次苏州，在质量方面并不逊色。南京亦有两座杰出园林。其他城镇，如常州、泰州、上海、南翔(上海郊区)、无锡等，均拥有古今闻名之园，值得一访。东南一隅之市镇，庙宇产业，抑或官家寓所、私有宅邸，皆有些许园林。人们常将杭州整座城市，视为一座围绕西湖之广袤园林，盖为所有城市中之最大水景。上海郊邻城镇，如昆山、松江、青浦、嘉定、江阴、太仓、吴江、常熟，更南之运河城市，如嘉兴、南浔、平湖及海盐，在其漫长历史进程中，皆因园林而添色。这些园林或幸存至今，或遗于坍毁而相继消失。除北方之北京、南方之广东等地区外，数量众多、不胜枚举之东南园墅，以园林城市苏州为典范，确立中国公共与私家园林之建筑风格。

四十多年前，两位著名纽约建筑师莅临上海，开启中国之旅，将苏州园林列为必访之地：一九三五年，伊莱·雅克·康[1]；一年后，克拉伦斯·斯坦因[2]及其妻，女演员艾琳·麦克马洪[3]。每及相伴共游之时，愉悦之情，难以言表。噫乎，余尚未讲解说明，彼即能针对中国景观艺术之独特审美因素做出本能反应，令人惊讶。每次出行均选在紫藤盛开之季，每日游程皆为一次完美之旅。

本书侧重园林植物配置方面，较诸其他地区，苏州园林拥有更为完整之类别。拟稿时，余时常查阅克里斯托弗·滕纳德[4]《现代景观中的园林》(Gardens in the Modern Landscape)，以及刘敦桢《苏州古典园林》，于此致以诚挚谢忱。

1. 伊莱·雅克·康(Ely Jacques Kahn，一八八四至一九七三年)，美国著名建筑师，曾在纽约设计多幢高层建筑。童寯、陈植曾经在其事务所工作。——译者注，余同。
2. 克拉伦斯·斯坦因，一八八二至一九七五年)，美国城市规划师、建筑师，是二十世纪初美国田园城市运动的代表人物。
3. 艾琳·麦克马洪(Aline Laveen Mac-Mahon，一八九九至一九九一年)，美国著名电影演员，在电影和电视界非常活跃，由于在《龙种》中的精彩表演曾获奥斯卡最佳女配角提名。
4. 克里斯托弗·滕纳德(Christopher Tunnard，一九一〇至一九七九年)，加拿大景观建筑师、城市规划师。

童寯
南京工学院建筑研究所
南京，一九八二年一月

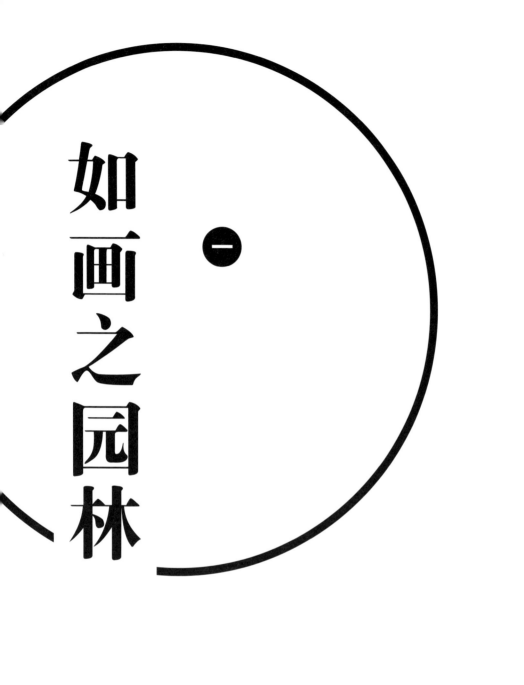

如画之园林

一

每展中国山水画卷，凡人鲜有纠葛，如此微小斗室茅舍，何以能入硕高之人？一道曲折蜿蜒之弱径，若干横跨湍流之薄板，竟将驴背沉醉之隐者，安稳驮载彼岸？唯览山水画卷而欲得观赏之妙，须先辨识赏画之反常规则。如此不合情理之格制，同样适用中国古典之园林。如是观之，彼实为三维中国画矣。

游者每探中国园林，甫入门园，徘徊未远，必先事停足。片刻踌躇实为明智，正因此行犹如探险。透过空间与体量，一瞥之下，全景转为一幅消除景深之平面，游者将深感惊奇，园林竟如此酷似山水画。虽非画家笔墨所绘，展陈游者眼前之无边景色，实则茅屋、溪流与垂柳等景物所构成，游者必能欣然联想与山水绘画之肖似程式——同样曲径，折向相似洞穴。比照之下，游者心满意足，怡然而去，将未赏之景，未睹之奇，留待日后再寻探访。正因为此，往昔园主鲜有朝夕居于园内者，偶至一访，保持相对距离，以添萦绕之魅力。

旧派论园者，以为绝妙佳园，必由丹青高手施以妙计。两世纪前，英国学者威廉·申斯通[1]亦有此论。申斯通断言，风景画家必为园林之最佳设计师。唐代王维(七

---

1. 威廉·申斯通(William Shenstone,一七一四至一七六三年)，英国诗人，业余园艺家和收藏家。申斯通继承了家族的李骚斯(Leasowes)庄园，并进行改造，使之从一座毫无特征的奶牛场转变成为一个拥有大片地坪、湖泊、溪流的著名景区，以对应古典诗歌中所描绘的特定场景。他在庄园周边设置一条环路，每隔一段距离设置座位，以便游客欣赏他精心培育的景观。他在李骚斯步道中使用许多铭文、诗歌和民谣，以唤起人们对于美丽风景的关注。

〇一至七六一年），诗画双绝，可为佳证。摩诘自营辋川别业，以养山居晚年。后世学者称王维"诗中有画，画中有诗"[1]，实先于勒内·德·吉拉尔丹[2]之论，"诗人感觉，画家慧眼"，长达一千年。王维无愧此誉。

绘画与园林，一如画师与造园家，二者密切维系，难分彼此。亚历山大·蒲柏[3]云"一切园艺皆绘事"，无意中将肯特与王维同称"绘画造园家"，兼画家、设计师、园主于一身。为摹写园中之景色，王维留传数幅绝妙绘画，竟成后世文人山水画之蓝本，造园之范式。王维其园其画皆为逸品，无可仿效，无可超越。托马斯·惠特利[4]则出语激进，竟称风景园并非"等同"而是"高于"风景绘画。

---

1. 来自苏轼《书摩诘〈蓝田烟雨图〉》，苏轼评王维语："味摩诘之诗，诗中有画；观摩诘之画，画中有诗。"
2. 勒内·德·吉拉尔丹(René de Girardin，一七三五至一八〇八年)，让·雅各·卢梭最后的学生。受卢梭的启发，创建法国第一座风景园林。吉拉尔丹曾出版《论风景的构成》(On the Composition of Landscapes)一书，认为风景对于人的感觉与灵魂有很大净化作用，如何去构造风景，事关国家在道德精神方面的振兴。
3. 亚历山大·蒲柏(Alexander Pope，一六八八至一七四四年)，英国十八世纪前期最著名诗人。他在文学上推崇新古典主义，讲法则，重节制。蒲柏是自然式园林的积极推动者，认为"没有装饰过的自然的那种和悦之简朴，在人的心灵上展开一片雅洁的宁静和愉快的崇高感"。
4. 托马斯·惠特利(Thomas Whately，一七二六至一七七二年)，英国政治家与作家，曾于一七六一至一七六八年担任英国国会议员、财政部长。作为园艺家，惠特利著有《现代造园艺术》(Observations on Modern Gardening)，该书曾是关于英国自然风光造园理论与方法最为全面的论著。

◇ 拙政园，别有洞天。展陈游者眼前之无边景色，实则茅屋、曲径与垂柳等景物所构成。

一　如画之园林　｜　27

◇ 环秀山庄，假山。中国古典之园林，实为三维中国画卷。

一 如画之园林 | 29

◇ 留园,曲溪楼。往昔园主鲜有朝夕居于园内者,偶至一访,保持相对距离,以添萦绕之魅力。

◇ 拙政园,梧竹幽居。风景园并非「等同」而是「高于」风景绘画。(左页图)

◇ 拙政园，补园折廊。诗中有画，画中有诗。

◇ 留园，水景。风景画家必为园林之最佳设计师。（左页图）

园林与文人

二

往昔中国园主，皆以文人园林为范本。富绅新贵，苦心经营；城市别墅，乡野庄园，竞相附会风雅，重品味而耻奢华。如此"显然无用"之园，以"城市山林"之局，为其主人提供摆脱尘嚣俗累之庇所，亦可铺陈园主于文化社会之角色，闲暇阶层之地位。即便一国之君，虽具权势，亦时有意欲从宫廷大内逃逸而出，以求在某处皇家园林别业内，得享片刻犹如乡绅之闲适生活。

居于此间，园主常取一"别号"，以假想自己为另一人物，如王维，如诗人，如画家，如隐者。

十七世纪之法国，情形迥异。法王路易十四之凡尔赛宫，非为隐居或冥思，而事炫耀与欢宴。相较凡尔赛宫之宏伟壮丽，世界其他宫殿皆显局促，路易十四亦不因隐避乡间而遭世人责难。与之相反，中国园林，无论皇室抑或私家，摩肩接踵不仅不合时宜，而且难以实现。其场合氛围，静谧清闲。通行之径，蜿蜒曲折，不容拥喧。

中国园林之另一特征，乃与文学之密切关联。诗人词家，文士墨客，名书镌刻，楹联匾额，如缺之，园中建筑则难称完美。各类题铭，须兼具华采之辞章，隽秀之书法，常悬于厅堂、亭榭或门道之上。每一单体建筑，常冠以得体合宜之雅称。

十八世纪之英国，亦有类似情形。诗人兼园师威廉·申斯通，于其庄园李骚斯[1]匾额之上，铭刻相应诗句，以表对景色之情思。申斯通为浪漫主义运动之魁首，影响深远，其追随者竟将如画之园引申为"每一小品，皆须题铭"。此者相应表明，铭文必会激发某种文学情思，引导游者通达视觉欣悦与哲学冥想。倘若中国园林高于绘画，倘若它是诗，或如小泉八云[2]所云，是幻想，其间点缀之题铭，恰好可以实现文人之理念，寓诗性想象于心灵之间。

唯文人，而非园艺学家或景观建筑师，才能因势利导，筹谋一座中国古典园林。即便一名业余爱好者，虽无盛名，若具勉可堪用之情趣，亦可完成这一诗性浪漫之使命。须记之，情趣在此之重要，远甚技巧与方法。

于文人而言，中国园林实为现世之梦幻虚境，臆造之浓缩世界，堪称虚拟艺术。吾人不必由此认为，游者全然愿意上当受骗。当其不再耽迷于园林，而开始生活于画卷之中，有关真实性问题则会迎刃而解。倘若某位东方哲人，于一幅画面中不能进入这亭那山而未觉忧恼，亦不会认为，

---

1. 位于英格兰西米德兰什罗普郡，面积约为二十三公顷，包含有别墅与风景园林。原属申斯通家族所有，申斯通逝后被爱德华·霍恩（Edward Horne）购买。
2. 小泉八云（一八五〇至一九〇四年），日本作家，出生于希腊，原名帕特里克·拉夫卡迪奥·赫恩（Patrick Lafcadio Hearn），一八九六年归化日本，改名小泉八云。小泉八云精通英语、法语、希腊语、西班牙语、拉丁语、希伯来语等多种语言，学识渊博。后半生致力推进东西文化交流，译作和介绍性文字很多，在促进不同文明的相互交流方面贡献非凡。

园林需要另当别论。再者，中国传统绘画，远非写实。丹青高手所绘，非其所见，而为其所想。

十六世纪前，日本园林全无可供步行之径，令人惊奇，难为现代心理所解。"可游之园"，须待后期方有。自曲廊之某一固定视点，观赏如画之景色，已令游者心满意足。倘若难以令人因其心灵之眼而去观赏一所无径之园，可以想见，今日西学之景观建筑师，若其业主要求设计一所完全无水无木之园，必感手足无措。日本园艺家具有设计"枯山水"之能，疏池理岸，营泉造瀑，不曾援用一滴水。中国园林设计之要义，须用想象，以谋小中见大，实中有虚。佛教禅宗引发日本枯山水，此种禅宗学说与威廉·布莱克[1]之"一沙一世界，一花一天堂"，于概念上异曲同工。

另一东方观念要求，观者须善于辨识因反差对比所致之景观，并接受罗伯特·福琼[2]之观点。福琼曾于一八四二年造访中国，收集植物标本。其推断，若要理解中国造园方略，须知"小中显大，大中见小"之技法。诚然，宇宙无垠之广袤，何种园林，任其扩

---

1. 威廉·布莱克(William Blake，一七五七至一八二七年)，英国诗人、水彩画家、版画家，英国文学史上最重要的伟大诗人之一，虔诚的基督教徒。早期作品简洁明快，中后期作品趋向玄妙深沉，充满神秘色彩。主要诗作有《纯真之歌》《经验之歌》等。
2. 罗伯特·福琼(Robert Fortune，一八一二至一八八〇年)，英国园艺家。曾受英国皇家园艺协会派遣，于一八三九至一八六〇年间四次来华调查及引种，发明并完善了长途运输植物的技术。他为英国皇家园艺协会引种野生或栽培的观赏植物及经济植物，收集花园、农业和气象情报资料，他还专门收集了北京故宫御花园中桃的栽培品种，不同品质的茶叶，收集荷花变种、佛手、金柑、食用百合及做宣纸的原料植物，分析植被茂密处自然土壤的理化性质。罗伯特·福琼的活动导致了中国制茶行业的衰退，使中国最重要的出口产品的贸易大幅度下降，严重打击了中国经济，改变了中国在世界经济中的地位。

展,皆仅为自然之微缩模仿。塞缪尔·约翰逊[1]于此表现极大隽智,并与中国(及日本)之无为哲学相契。约翰逊以为:"略具一二即佳。"为扩大园之域野,园林入口刻意采用反差,力求不甚显眼,低调寻常,以便访客不拘礼节,犹如悄然潜入。与之相反,欧洲园墅之大门,造作虚张,浮夸修饰,遂使东方人于出园间,难获一枕黄粱之快感,进而难免疑虑,此去是否正在回归自然。

---

1. 塞缪尔·约翰逊(Samuel Johnson,一七〇九至一七八四年),英国诗人、评论家、散文家与辞典编纂者,英国十八世纪中叶以后的文坛领袖。他不仅以其作品称著,而且谈吐机智隽永,刚劲有力,在整个英国文学界其知名度,仅次于莎士比亚,并被人引用得最多。约翰逊编纂的《英文辞典》对英语发展做出了重大贡献,开创了英文词典学的新阶段。约翰逊从大量文学著作中搜集素材,选出例词例句,同时还注意日常用词的解释,并对当时的英文拼法起了规范作用。在一八二八年美国《韦氏大词典》问世前,它是最具权威的英文词典,并受到法国和意大利学者的赞赏。

◇ 网师园,真意门洞。情趣在此之重要,远甚技巧与方法。(右页图)

◇ 艺圃,南侧假山。城市山林。

◇ 拙政园，玉兰堂。诗人词家，文士墨客，名书镌刻，楹联匾额，如缺之，园中建筑则难称完美。

◇ 留园，涵碧山房。其场合氛围，静谧清闲。通行之径，蜿蜒曲折，不容拥喧。（左页图）

二　园林与文人 | 43

建筑与布局 (三)

一座缺少建筑之中国园林，确实难得。饶有情趣、恰当布局之建筑，诚为一座优美园林之首要。法国风景画师法雷斯特[1]亦有此论。其断言："一座优美园林附带一所优美建筑，实属罕见。"

无所不在，但不能忽略之建筑，当属凉亭。此构筑犹如玩物，甚至可伫于独柱之上。平面取形三角、单圆、任意多边，或双圆，或十字，由此，其屋顶形式可获从单一攒尖至多尖、多坡之各种类型、各种组合。反翘檐口，曲线屋脊，以及其他建筑不规则之做法，可为园林已有鲜活情调，更添不少佳趣。

一跨或数跨大型结构，列柱所构之崇厅高堂，踞于要冲，成为园林构图之一道强音。可拆卸之窗牖，环四面设置，皆可任意打开，以骋目放望，观景赏色。大型厅堂往往于其正面，附有升高平台或宽敞铺地。然而书斋，则应设处幽僻场所。

带顶游廊为另一园林特色建筑，偶有双层，实为柱廊，以联络各式建筑。廊或处两座庭院之间，因其开敞与透彻，而获既分犹合之效。如欲两院全然隔离，只须将游廊以壁屏之。一旦游廊浮于水面，则会采取带顶桥廊。

1. 让·C. N. 法雷斯特（Jean C. N. Forestier，一八六一至一九三〇年），法国景观建筑师，倡导保护巴黎的步行环境，曾设计埃菲尔铁塔下方的战神广场（Champ-de-Mars）。

长廊并非平直通彻，常呈之折或波曲。如在山林地，须随地势曲折婉转，顺坡面起伏成阶。一座园林建筑或可作为对景或观景中心，每有树木、花卉，或其他装饰陪衬，尤为如此。鉴于成组之建筑，游廊位置取于建筑之相对重要处，并相应决定其间之空间。此种布局须要考虑韵律和谐。谨记一种弊端：过冗建筑，杂而无序，势将导致沉闷或幽闭。

园林建筑，亦有类如英式"小品"之石船或旱舫。模拟舟楫，实为有趣。倚靠近岸，静泊湖面，游者于不动舱中，透出窗牖，观赏湖面景色。石铺舱面，对弈、品茗之所。

园林可建山地斜坡之上。中国营园者，于此发挥建筑布局之伟大创造力，与意大利人经营系列层叠台地，有异曲同工之妙。中国园匠善用台地，于其下方设置幽室，其上则可用于种植或漫步。一方台地，亦可成为一座庭院，以致幽隐。再者，台地高度亦可利用，以助俯瞰相邻较低之园林景色，眺望周围有庙有塔之郊外远景，因而曾被誉为"借景"。在此，园林如画之境界，似得倍增。营园者如获机遇，则最喜采用该类主题。

日文"しゃっけい"与中文之借景，文字对应，词义

相同。令人不由想起，从波波利花园[1]远眺伯鲁乃列斯基穹顶[2]之动人意象，从美第奇别墅[3]平台喷泉之后方，观赏圣彼得大教堂穹顶之壮美风景。

一湾碧池，尺度妙巧，鸳鸯悠游，锦鲤潜行，诚为园林欲求之景，似无一例外。与此相应，摩尔人[4]于园林中，亦以天鹅隐喻幸福，以白鹤象征长寿，作为隐居之伴侣。日本从中国引入这类因素，用以活跃园林氛围。至于荷花，应对生长区域加以限制，留出水面空间，形成倒影。湖中孤岛，一叶轻舟，亦可构成动人景观。池边可用泥砂、毛石、片岩砌筑，或以叠石点缀。

---

1. 波波利花园（Boboli Gardens），位于意大利佛罗伦萨，美第奇家族在一五四九年买下皮蒂宫后建造，是一座典型的文艺复兴风格园林，一七六六年向民众开放。
2. 指佛罗伦萨圣母百花大教堂穹顶，由文艺复兴时期建筑师伯鲁乃列斯基设计。
3. 美第奇别墅（Villa Medici）位于意大利罗马，是西班牙广场上方平乔山顶的一组建筑群。现为法国学院所在地。美第奇别墅前方的平台因地势较高，可以眺望罗马城市全景。
4. 摩尔人是于中世纪时期居住在伊比利亚半岛（今西班牙和葡萄牙）、西西里岛、马耳他、马格里布和西非的穆斯林，西班牙人与柏柏尔人的混血后代，他们于十一至十七世纪创造了阿拉伯、安达卢西亚文化。

◇ 留园，涵碧山房。一座优美园林附带一所优美建筑，实属罕见。

◇ 拙政园，枇杷园。但不能忽略之建筑，当属凉亭。

◇ 拙政园，与谁同坐轩。如在山林地，须随地势曲折婉转，顺坡面起伏成阶。

四 装修与家具

中式园林常环以高垣，犹如欧洲中世纪庭园，屏蔽凡俗之外界。各院落间如未隔以建筑，两侧则设廊墙，以为间隔，既因功能所需，又为装饰所用。墙垣绝非光整而平常之砖砌体，常有平面弯曲且端顶起伏者，并饰以奇思妙想之饰物。壁面常零散镶嵌书法刻石，或以薄砖、片瓦、砌构带有精美图案之格栅漏窗，以消除单调空白。日光投射，漏窗因其深度而呈闪烁景象，增添媚人之处。再者，同一漏窗于不同角度光照之下，亦呈莫测变化。此外，若以摩尔人视角观之，自墙垣空框内，可窥别院之只角片景，如此，则将空间延伸更远。正如现代建筑之拼贴，此类墙洞可使人观察两侧风景，重叠错落，穿插并致，可谓园林艺术之拼贴手法。

凉亭或花架，皆可倚墙置。如在园中，其可倏然而止，并续以假山石屏。再者，墙面刷白，充当"映印"植物(如竹)阴影之用，或作为一峰叠石、一棵奇树之背景。白粉之墙、灰瓦之顶，绿树浓荫，涂漆木构，决定园林之主体色调。中国园林，注重含蓄，多以折墙，掩映其美，诱使游者透过门洞或漏窗，进行窥视。门洞常以满月、宝瓶或花瓣为形，如同漏窗，其造型方式无可穷尽。

中国园林之铺地，亦为一种高超装饰之艺术。常以寻常之物，甚至废弃材料而倍添趣味。碎石、断瓦、卵石以及瓷器残片所构镶嵌纹理，于构图与色彩方面样式多变。此类设计通常采用对称、四边或其他多边等形式进行组合。非对称形式，一般以"冰裂纹"式样为特征。西国偶有案例，如意大利城镇埃斯泰之别墅[1]，其铺地于气质而非外观上，与中式铺地多有惊人相似之处。

园之中，室外、室内之家具，虽为末节，然其实用装饰，最不可略。庭园之桌椅，主要由雕琢石材或其他耐久材料制成。内部陈设，常用精选硬木。悬于顶篷之灯具，用于照明，亦为装饰。值得存疑之处在于，中国园林从未考虑夜间地面之照明。日本石灯，原为中国寺庙之附物，却于中国庭园中未获采用，实为意外。

为寻求变化，增添情趣，园林之浅盘缩景，盆栽罐植，错落配置。沿水边置护栏，多用精致木雕或石琢之材，用以修饰凉亭和平台。但最具价值之园林装饰，无疑当属叠石。

---

1. 埃斯泰别墅(Villa d'Este)，位于意大利罗马的东郊小镇蒂沃利，距罗马二十多千米。因为园中有许多漂亮的喷泉，又称"千泉宫"。

◇ 网师园。若以摩尔人视角观之，自墙垣空框内，可窥别院之只角片景，如此，则将空间延伸更远。

◇ 留园,铺地。中国园林之铺地,亦为一种高超装饰之艺术。

◇ 留园,花步小筑。墙面刷白,充当『映印』植物(如竹)阴影之用。(左页图)

◇ 网师园。内部陈设，常用精选硬木。

◇ 网师园。

◇ 网师园。

◇ 网师园。

四　装修与家具　| 59

◇艺圃

叠石 五

传统中国园林，假山为最奇特之物。盖具半自然、半人工之特征，假山常置人工构筑与自然种植之间，于静态建筑与脉动青翠之中，架起一道令人赞叹之桥梁。然而此处之假山，必然异于罗马别墅原址之巨石，西塞罗[1]曾偶遇发现，并在其文章中大加赞赏。亦不同于十八世纪惠特利于英国风景园林中所指用于装饰之石材。（萨里[2]之凝灰岩拱门与洞穴理应排除在外。）

叠石大都自远方获至，甚至距其最终目的地远至数百千米。价格最昂贵者，应属经水力冲蚀而成之"湖石"。湖石采自深水，历经数世纪强烈水流之持续冲击与洗刷，穿凿而成孔洞，表面凹凸起皱。苏州邻近太湖，这一园林城市因此聚集大量可用之材，石工技巧且经济，城市本身亦成著名之观览圣地。

湖石经由湖水浸润冲刷，多孔而奇异，极富特质。由于抽象轮廓及虚实体量，湖石极似亨利·摩尔[3]或野口勇[4]之现代雕塑，令人匪夷所思。如湖石仅有一峰，可如欧洲雕塑，置于座上，成为一种独特装饰；如湖石不止一峰，石块可粘接成形，以构造洞穴、蹬道、山

1. 马库斯·图留斯·西塞罗(Marcus Tullius Cicero，公元前一〇六至公元前四三年)，古罗马著名政治家、演说家、雄辩家、法学家和哲学家。
2. 萨里(Surrey)，英格兰东南部行政郡和历史郡，位于伦敦西南四十八千米处，邻近泰晤士河。面积一千六百六十三平方千米，风景优美。
3. 亨利·摩尔(Henry Moore，一八九八至一九八六年)，英国雕塑家，其作品经常采用一种半抽象化方式进行创作，因而发展了一种新风格。
4. 野口勇(Isamu Noguchi，一九〇四至一九八八年)，日裔美国雕塑家和设计家，二十世纪最著名、最具批判性的学院派雕塑家之一，最早尝试将雕塑和景观设计结合。

峰。某些叠石如此庞大，以至占据大部分空间而成园林主题，如扬州著名历史园林"万石园"[1]。现存苏州"狮子林"则为另一案例。是园影响广泛，但亦有几分杂乱，曾不幸被贬为"乱堆煤渣"[2]。

昔日中国叠石鉴赏家，曾编撰多册《石谱》。美国加州大学伯克利分校于一九六一年重印杜绾之《云林石谱》[3]，将其视为一篇宝贵文献。该图谱由杰出艺术家绘制、镌刻，并加以学术性注释，以图像描绘不同时期、不同地区名石之显著特征。杜绾作石谱列举一百一十六种名石。旧时文人对于名石之评价标准，可归为四条：漏、透、瘦、皱。其中所蕴含之情感，如非源于湖石之抽象美学，则必定来自神秘主义。石有"品格"，因而致使文人爱石之嗜好，近乎发狂程度。其因在于石之恒久、不化、坚定之特征，而此恰为人类品质所经常欠缺。噫！或许莎士比亚不幸错言，因其曾经借用安东尼之演说，否认石之智慧与感觉。

石虽无言，却因金子般之沉默而得文人之敬仰，并被视为一种最宜人之品格（石不能言最可人）。

就中国文人之石癖，甚至叠石崇拜而论，莎翁同样有

---

1. 《（嘉庆）重修扬州府志》记载园中假山为石涛所作。王振世所著《扬州览胜录》也有提及。万石园原属扬州余氏。
2. 沈复在《浮生六记》中的评语。
3. 杜绾所撰《云林石谱》书成于南宋绍兴三年（一一三三年），是中国最早、最全、最有价值的石谱。其中涉及各种名石一百一十六种，对其生产之地、采取之法，以及形状、色泽有详细表述。杜绾，字季阳，号"云林居士"，浙江山阴人，平生好石。其祖父杜衍北宋庆历年间为相，封祁国公，父亲也为朝中重臣，姑父是著名文学家苏舜钦。

所简化。莎士比亚时代之前四百年,一位中国文人仕官曾经豪放不羁,竟与一块引其想入非非之顽石称兄道弟,但也招致其仕途难以为继[1]。数百年后,另一文人热衷一峰异常秀美之假山,又有一人却如此着魔一枚微小玩石,难舍难分,竟致无此物为伴,不能安卧入眠[2]。

传统园林中,叠石之作,极为精巧,难度最大,已成共识。历史中,仅有少数"匠师"可以运用自如。

十七世纪张南垣[3]为最著名叠石大家,不屑以一堆石块,模拟自然山峰,而关注以有机随意、不落筌蹄之构思,突出天然要素之本质。仅以少许叠石之工作,喻示峦嶂之神态,并大获其成。

一世纪后,有戈裕良[4]者,倾其辛劳,营造石山。规模常属中等,然颇为杰出,意味风雅而神采超逸。今日苏州小园环秀山庄,尚存戈裕良一例无与伦比之作品,可资瞻赏。戈裕良于此之功,在于针对洞穴顶部构造进行优化,以石拱砌筑替代历史悠久之平板铺设。当时及后代文人认为,戈裕良实为叠石造山之第一宗匠。

历史所载最早石山,可溯至公元前一世纪,但极为重要、规模最大之叠山,当数宋

1. 指宋代米芾。《梁溪漫志》曾载:"米元章守濡须,闻有怪石在河壖,莫知其所由来。人以为异而不敢取。公命移至州治,为燕游之玩。石至而惊,遽命设席,拜于庭下曰:'吾欲见石兄二十年矣。'言者以为罪,坐是罢去。"
2. 指明代米万钟。米万钟爱石成癖,时称"友石先生"。
3. 张南垣(一五八七至一六七一年),名涟,字南垣,以字行。明末清初造园家。
4. 戈裕良(一七六四至一八三〇年),字立三,年少时造园叠山,曾创"钩带法",使假山浑然一体。

代徽宗皇帝[1]之古代皇家石园。徽宗本人之画艺，举世无双。其耗费过多心智于叠山营园，而非国家大事。大运河畔，舳舻云集，太湖之石，满载而至，运达宫内，营建徽宗所醉心之石园"艮岳"[2]（东北顶峰）。某日，来自北方之女真，围攻都城。靖康蒙难，国破家亡，大量假山名石，不得已碎作石弹，以卫城郭。

女真人精选一些名石，运至中都（今北京），以饰皇家花园之岛中小山，即今日之北海琼华岛。

---

1. 宋代徽宗皇帝，于一一〇〇至一一二六年在位，宋朝第八位皇帝，擅书画，他自创一种书法字体，被后人称为"瘦金体"，热爱画花鸟画，自成"院体"，是少有的艺术天才与全才。
2. 艮岳，中国宋代的著名宫苑。宋徽宗政和七年（一一一七年）兴工，宣和四年（一一二二年）竣工，初名万岁山，后改名艮岳，一一二七年金人攻陷汴京后被拆毁。艮为地处宫城东北隅之意。

◇ 太湖三山岛，湖石。湖石采自深水，历经数世纪强烈水流之持续冲击与洗刷，穿凿而成孔洞，表面凹凸起皱。

五　叠石 | 67

◇ 苏州振华女中，瑞云峰。传统中国园林，假山为最奇特之物。

◇ 留园，瑞云峰。湖石经由湖水浸润冲刷，多孔而奇异，极富特质。

◇ 环秀山庄,假山。规模常属中等,然颇为杰出,意味风雅而神采超逸。

◇ 耦园，黄石假山。仅以少许叠石之工作，喻示峦嶂之神态。

五　叠石　｜　71

◇ 狮子林。石虽无言，却因金子般之沉默而得文人之敬仰。

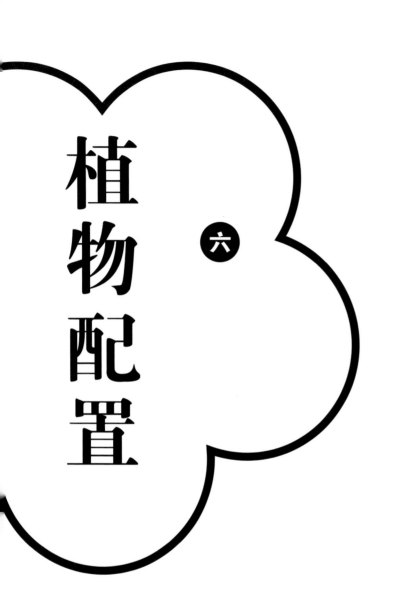

植物配置 六

昔日中国园林匠师，鲜有注重园林植物者，其因并非青翠事物难堪大任，而在于其固有观点：花木仅起附属与辅助之用。

日本园林之匠师，因相同缘由，将花卉树木减至最少，不足为怪。桃李竹柳，松柏桂梅，兰蕙夹竹，红蓼紫葵，欲以有机方式，调谐环境。除地形、日照、湿度、生长习性，以及本地植物环境特征外，姿态、纹样、外观、轮廓、色彩与芳香，构成如何挑选植物品种之基本准则。

相比种植因素，植物之生命特征更为重要，不仅可供缓解一览无余之尴尬，亦可为诗情画意提供适宜素材。如松、竹、梅，即通常所谓"岁寒三友"，即为常规属类。古树嘉木，历来都因饱经沧桑、永不屈服而备受褒扬，并于园林之旧址新建中，广受欢迎。

中国园林虽为一种亲切、宜人、注重经验之艺术，但植物生长通常不显人工斧痕，此理须识。中国园林，皆无布局整齐之绿篱，亦无几何图形之花坛，修杈剪枝必然摒除于外。无论欧洲园师如何巧于植物配置，殚精竭虑，中国园匠仍然执着于凉亭修饰，不惜工本之配景小品。

最明显差异在于，中国园林必不见有边界分明、修剪齐整之草坪，因其仅对奶牛颇具诱惑，实难打动人类心智。

风水意识以至迷信观念，往往可使现存地形连同附属景物维持不变。自然野趣，听之任之，此种审慎之觉，可使中国园林永不混同于西式植物园。

约翰逊博士承认无法分辨甘蓝玫瑰与甘蓝（俗称卷心菜）之别，却武断质问鲍斯韦尔[1]："难道园林不都是植物园吗？"此位辞典编纂大师之言，于英语角度毋庸置疑，但于中文用法上，"植物"一词常用于治疗疾病之药草，只与药剂师及病人相关。显然，鲍斯韦尔好友威廉·钱伯斯爵士[2]，曾经访问并仰慕中国园林，却未能向其传达东方艺术之精神本质。树木、花卉永远不会成为园林之特征，但却居有相应地位，抑或极为突出。

十二世纪洛阳之牡丹，十八世纪扬州之芍药，皆享极高声誉，所妆点之园林，因其获致传世盛名。即使如此，精心选择花木，仍须呈现有机而不宜唐突。英国风景园林受此衣钵，却走向极端。每当布朗[3]获得"改建"之机，经其"有为"之手，花卉竟然一洗逸尽。

倘若中国园林无须歌功颂德，树碑立

---

1. 詹姆斯·鲍斯韦尔(James Boswell, 一七四〇至一七九五年)，英国文学大师、传记作家、现代传记文学的开创者，出生于苏格兰贵族家庭，英国文坛领袖塞缪尔·约翰逊的挚友，并为其撰写传记。
2. 威廉·钱伯斯爵士(Sir William Chambers, 一七二三至一七九六年)，英国乔治时代最负盛名的建筑师，当时帕拉第奥式建筑的先导者之一。钱伯斯曾于十八世纪四十年代到过中国广州，对中国建筑、园林和工艺美术有极大兴趣。主持过邱园(Kew Garden)的设计，一七八二年任宫廷建筑师。一七五七年出版《中国建筑、家具、服装和器物的设计》(Designs of Chinese buildings, furniture, dresses, machines, and utensils)；一七七二年出版《论东方造园艺术》(A Dissertation on Oriental Gardening)。
3. 兰斯洛特·布朗(Lancelot Brown, 一七一五至一七八三年)，亦称万能布朗，英国最著名的园林设计师，擅长自然式园林设计，将园林与建筑融为一体。设计取材于自然，从不使用雕琢过的石料或建筑形体，通过开阔草地和平静水面，利用起伏地势，栽植树木，创造和谐精美的风景。

传，植物于此难有作为。中国园林中，历来无有开阔大道、圆形广场、高松阔柏、宏壮喷泉，凡此种种，似为王尔德[1]之所谓"自然个性之双重表象"。

再者，西方园林格局中之主轴与次轴，此类布置仅能导致敬畏、感化与悲壮。中国园林中，此类景观同样不可思议。西方园林从未摆脱树木辽阔之外观。据勒诺特尔[2]之惯例，青翠植物必然主导各式园林，倘若不加控制，任其发展，野草杂木势必四处蔓延，并最终回返"原始森林"之态。众所周知，勒诺特尔强调面向大自然之彻敞，而不能容忍阻视远景之事物。而此时，则应重新审视花木与庭院之人工造物之关系。

倘若庭园宽敞，古树巨木，当空伫立，可构一幅佳图，亦可用于协调不同建筑物之关系。此外，嘉树犹如枢轴，可用以限定主次庭院空间，其诗意名称，亦可用于题写一座建筑或一进院落之匾额。如遇环境狭窄、视线受制时，一株大树已无地可容，布局准则适宜调整，配以轮廓优美、色调丰富、芬芳宜人之灌丛，并可辅以一峰孤石。

混合种植，应以落叶乔木为主，并辅以常绿之嘉树或秀竹，再为攀缘植物或禾本植物所环绕。粗壮树干，苍劲

1. 奥斯卡·王尔德(Oscar Wilde，一八五四至一九〇〇年)，爱尔兰作家、诗人、戏剧家。十九世纪末英国唯美主义运动的主要代表，"为艺术而艺术"的倡导者，主张美没有功利主义价值，并且与道德无关。
2. 安德烈·勒诺特尔(Andre Le Notre，一六一三至一七〇〇年)，法国最负盛名的景观建筑师，法王路易十四首席御用园艺师，曾设计路易十四的全部园林和达官贵人们的私家花园，以及大量小型的花园。其作品除了在法国之外，还分布在德国、英国、比利时和意大利，总数达上百个。

枝杈，须避柔枝弱叶。大簇落叶之木可与常绿青树相配，以表稀疏与浓密之对比，浅色与深色之反差，并与季节共变，避免单调与重复。春季之玉兰与牡丹，夏季之紫薇与荷莲，秋季之菊花与枫树，蜡梅与茶花则盛放于冬季。一座园林庭院中，可观一年四季，各类花卉，交替盛放，诚为理想之境。

相较北方城市，苏州深得自然之垂宠，适宜生长各类树木植被。城中任何一座大型园林，皆可列举一百余种植物，中等及小型园林，亦可多达二十至七十余种。荷花、紫藤、梅、桂、秋海棠、茉莉和菊花，各园俱有。如若缺少竹林，无园堪称完满。青松、翠柏与绿杉，常绿树是也。柳木、红枫、梧桐、棕榈、芭蕉、榆树，应有尽有。自古以来，树木花卉之移植与杂交育种，始终有之。中国园林之苗木栽培，技艺精巧，各类植物，可资植物学科学家撰写一部专著。

根据视觉体验与实用价值，园林植物大致可分以下种类：

1.观花类。园林中最赏心悦目之事，即为花卉植物之色彩或芳香，兼而有之更佳。常绿类植物中，有山茶、玉兰、月季、桂花、杜鹃、夹竹桃、栀子花、六月雪、含笑等。落叶类则有牡丹、梅、桃树、杏、海棠、紫丁香、芙蓉等。牡丹因其辉煌夺目之色彩，丰硕非凡之尺寸，于所有园林花坛之上，无疑为最端庄者。每至寒冬季节，其他植物凋零败落之时，山茶、梅花尤受欢迎。

2.观果类。自初夏至深秋，多种林木开始结果，色彩纷呈且多数可食，如金色之枇杷、橙黄之香橼、南天竹，无花果及枣树则为落叶类。

3.观叶类。无论何种园林，此类植物都难以或缺。可资全年妆点之用者，有黄杨、丝兰与棕榈。垂柳、柽柳与枫树，于冬季木叶凋零；其余季节，仍为丛林青翠增添色彩，变换姿态。尤以枫树，其绯红光彩，可于深秋之日，构造非凡之特效。

4.另有树木，实质为种植之基本构成，于炎热夏季，可供遮阴。诸如桧柏等针叶树，与落叶类之银杏、榆木、洋槐，尤其与叶茂枝繁之枫杨组合一起，可营造锦团花簇、落英缤纷之氛围。

切勿遗忘，尚有攀缘类，如常绿类之蔷薇与常春藤，落叶类之葡萄藤、紫藤与牵牛花，均可用于攀爬棚架，添加装饰，填充空隙，用以遮掩建筑或假山之光秃表面。另有植物，可增添聆听之悦，如穿梭风中秋叶之瑟鸣声，滴打水中睡莲之雨点声,以及树草鸟虫之吱喳嘀啾。除此以外,另有多种竹科、草类生长物,辅以荷花、睡莲等水中植物,后者围以芦苇,逼涌水岸,将水沼之感传给周遭环境。

◇ 狮子林,银杏。古树嘉木,历来都因饱经沧桑、永不屈服而备受褒扬。(右页图)

◇ 拙政园。中国园林必不见有边界分明、修剪齐整之草坪。

◇ 网师园,凌霄。(右页图)

◇ 网师园,玉兰。

◇ 网师园，紫藤。

◇ 网师园，芭蕉。

◇ 留园。

东西方比较（七）

法国某诗人曾云"吾甚爱野趣横生之园"[1]。此语恰好表明西方园林与中国园林之区别，皆因后者全然摈弃山野丛林之气。中国园林实非某一整体之开敞空间，而由廊道与墙垣分隔，成若干庭院。主导景观并成观者视觉之焦点者，为建筑而非植物。中国园林中，建筑如此赏心悦目，鲜活成趣，令人轻松愉悦，即便无有花卉树木，依然成为园林。此在日本尤为真切。京都龙安寺园内，完全摒绝植物，只现石、砂，以及一道夯捣土墙。该园借用紧接边界外之茂密树林，以资弥补，保持高雅。

西方园林截然相异，其地景远甚建筑，以致建筑有如孤岛置汪洋中。林木、花卉及喷泉，相较别庄与凉亭，彼此关系更为紧密。模拟建筑以塑形，遵从轴线以布局。

认真研究中国园林之英国学者，威廉·钱伯斯实为第一人。彼于著作《论东方造园艺术》中，力图实证中国风景艺术之优越。钱伯斯有幸于中国园林艺术之鼎盛时期，亦即清代乾隆年间（一七三六至一七九五年），造访中国。其主要论点为：对于中国园林与欧式园林，拘泥相对优点，争论孰优孰劣，实为徒劳；倘若与其各自领域中之相关艺术、哲学及生活和谐一致，则两者同样伟大。

---

1. 法国诗人皮埃尔·德龙沙（Pierre de Ronsard，一五二四至一五八五年）语。

格拉纳达之阿尔罕布拉宫[1]，介于东西方之间。这一伊斯兰城堡，汇集众多厅堂，庭院、游廊环绕聚之，虽颇合中国风范，绿圃水池却按几何样式设计。尽管对称原则无处不在，但布局宜人随和，无西方园林之刻板与单调。

意大利式别墅，受惠于秀美地形与温和气候，尤擅经营台地、茂盛植物，影响波及西方古典园林。花坛、雕像、阶梯、瀑布，都以规整格式布局，列序井然。高大柏树，排阵森严。罗马花园，善用轴线与重复之手法，并借助层层叠台，确实有助达到引发惊喜这一目的，这恰与中国园林设计原理不谋而合。攀登意大利式台地别墅之梯级，驻步平台，俯视眺望，一片无与伦比之壮丽景色，呈现眼前。

几乎所有中国园林庭院，皆助访者获此相似体验。然而，由此领略并非一瞥中之全景。眼前所见，仅为整体之某一局部。每当游者向前探寻，一次又一次，常会深感意外，噫嘻太息。眼前世界，逐次展开，诗句铭文，激发想象，无边景色，引人遐思。法兰西为注重理性、强调逻辑之国，园中意外因素，理应摒除。平坦地形，辽阔尺度，于公园中，则进而强化此种单调感。

诚然，中国园林亦有不尽合理之处。以西方思维方式观之，必然可笑。一座亭阁之

---

1. 阿尔罕布拉宫是中世纪摩尔人在西班牙建立的格拉纳达埃米尔国的王宫，有"宫殿之城"和"世界奇迹"之称。格拉纳达（Granada）是西班牙安达卢西亚地区格拉纳达省的省会，著名历史城市，融汇伊斯兰教、犹太教和基督教风格，是西班牙的文化和旅游热点。

楼层，竟似无径可入！若获可登之梯，实属幸事。蜿蜒狭窄之步道，于两点之间，却须穿行颇远距离，令人不得其解。再者，光滑、陡峭之石山，如此险危，其意在于劝阻攀爬。一条蜿蜒逶迤小溪，一道低矮曲折石桥，其意并非引导游园者顺利跨越，而似引导其濒临淹溺其中，实为奇特。相较惶惑，困顿更为糟糕！欧式园林之游者，深谙意大利式别墅之端庄，熟知英伦风景之魅力，然甫入中国园林，对其率性与放诞必感手足无措。

不期而至，未可预测，此类特征皆属一类思想派别，与中国文人哲学完全合拍。然而，肯特[1]有关"大自然不喜直线"之断言，非与这一中国哲学不谋而合欤？倘若笔直行道，漫长阔路，规则花坛，诚为肯尼斯·克拉克[2]所谓西方数学观念中"心灵度量结构"之结果，中国园林，实为不能度量之艺术。可谓不惜任何代价，以避僵硬几何、刻板秩序。

园林之曲径，以其处心积虑、刻意斟酌之不规则性，可称为"sharawadgi"（曲折有致）[3]或"无秩序美"，成其

---

1. 威廉·肯特（William Kent，一六八五至一七四八年），英国画家和建筑家，他设计的花园崇尚自由、自然。
2. 肯尼斯·克拉克（Kenneth Mackenzie Clark，一九〇三至一九八三年），英国艺术史家，意大利文艺复兴艺术方面的权威学者。
3. 英国政治家兼作家威廉·坦普尔爵士（Sir William Temple，一六二八至一六九九年）在一六八五年所撰写的《论伊壁鸠鲁的花园》（Upon the Gardens of Epicurus）一文中，提及"sharawadgi"一词。该词是杜撰之词，意思是"千变万化"，或"诗情画意"。造园史家克里斯托夫·撒克尔（Christopher Thacker）认为，在十八世纪初，该词用来指称反几何式花园。在一九八九年版《牛津英语辞典》中，关于sharawaggi（也可写为sharawadgi）的相关解释为："一六八五年，威廉·坦普尔在《论伊壁鸠鲁的花园》一文中提到，中国人有一种特殊的词语，以表达一种经由精密推敲的不规则美，可以令人一见倾心。汉弗莱斯（A. R. Humphreys）认为sharawadgi在英国存在三种理解：不遵从数学原理，信奉不规则；在无穷变化之中寻找美；按照材料本身潜在的有机模式来对待自然材料。"

营造特征。中国园林空间处理，皆将观者视域局限于仅为单一画面之庭院中，旨在充分体现隐匿与探索之主题。游者于迷惑中，将始终乐此不疲。幻境迭生，迷津不断，所穷无尽。然变动不居，随性漫游，岂不比简单达到更具情趣乎？

对于率性而为、放诞不羁之艺术家而言，某种被延迟之快感，实为双倍享受之愉悦。园林犹如万花筒之构图，有助放松精神，随机漫行。游者时为各类趣味所吸引，通过开敞与封闭之空间对比，阴暗与明亮之场所对比，高岸与低平之孔洞对比，巨大与窄小之平面体量对比，错综之局，复杂之景，因对比而更为增强。

为创造万千变化之景色，不仅步道成曲折之形，地形标高往往亦成不规则起伏状，进而屏遮视线，将视觉感知限定控制于某一局部。凡尔赛花园与之相异，其开敞布局与广阔视野，委实令人乏味，以致不得不另建迷宫，添增游径。为消除线性单调感，林木灌丛间夹隐秘小品，以诱引好奇与随性之人。由此只可推断，西方园林实则悦目，中国园林意在会心，其一信奉量测标杆，另一追崇心智沟通。追根溯源，实为精神与物质之问题。

一经建成，中国园林于初期阶段，伴随建筑与其他人工物逐年成熟而日臻丰美，植物则延迟在后。经年后，时值林木苍古幽雅，建筑又近乎失修。坚固耐久之叠石，却能保持最为长久。中国传统园林中，这一问题最为恼人，却不如西方欧式园林，以建筑为重点。然东方哲学家却以超然沉着，以待如此盛衰交替之势。这一超然易于理解，因为园主只择最佳时刻，才事园林观赏，正如玩赏稀有古画。两者都须有待时机，年代愈久，价值越珍。

经由该篇论文可以得知，中国传统园林与十八世纪英国如画风景共有某类特征。倘若效仿即为一种直截了当之恭维，英国浪漫主义学派于自觉或不自觉间，以同一方式模拟中国范例，并致以最高敬意。

自威廉·钱伯斯以降，更多中国园林之外籍爱好者，为此著书。奥斯瓦尔德·喜仁龙[1]所撰《中国园林》，一九四九年出版，述其在中国北方所见之园林。南斯[2]久居苏州，一九三六年出版论文集《园林城市苏州》（上海凯利与沃尔什出版）。该书作者朴素平实，并非专为学术目的，亦为普及观光所撰。该册小书甚为可爱，所述苏州六座主要园林，均值一访，加以研究。塔尔博特·哈姆林[3]所著《二十世纪建筑的形式与功能》

1. 奥斯瓦尔德·喜仁龙（Osvald Siren，一八七九至一九六六年），瑞典艺术史学家，研究范围涉及文艺复兴时期的意大利、十八世纪的瑞典以及中国艺术。喜仁龙曾多次前来中国考察，著有《中国园林》《中国园林及其对欧洲园林的影响》《中国雕刻》《中国绘画史》《北京的城墙和城门》等。
2. 弗洛伦斯·拉什·南斯（Florence Rush Nance，一八六八至一九六八年），曾于一九三六年出版《园林城市苏州》。
3. 塔尔博特·哈姆林（Talbot Hamlin，一八八九至一九五六年），美国建筑师和建筑历史学家，曾担任哥伦比亚大学建筑学院教授。

（一九五二年），内有一章，专门讲述"园林与建筑"，读者可以觅得两张苏州园林平面图，以及针对曲折游径、各类视景、变幻地貌，神秘与高潮之叙述，所有内容皆为本质因素，均有助提升中国园林之梦幻如画之境界。中国园林偕同世界其他园林，皆为卓越平和之艺术。

劫掠、战祸与自然侵蚀，曾为园林之主要破坏因素。贫困潦倒、缺乏担当之园主，即便于和平时代，亦会轻易荒弃园林，任其衰败。战争时期，极少园林可以幸存而免于倾圮损毁。一九四九年后，中华人民共和国进行大量修缮与修复，尤其在苏州及其他地方，极大程度复兴了这一辉煌艺术。然而，吾人无论如何突出苏州，都将永不过分，该城拥有大量古典园林，著名与不太知著者，大小总数超出一百座。苏州由此于各城市中，确立无可动摇之地位。

于此之外，万切不可忘却，除暴力拆毁，尚有渐微平缓之力，亦即西方景观建筑学。这一当前于中国迅速成为各个学院之时髦课程，正在削弱中国古典园林世代相传却已危如累卵之基础。倘若怠懈放任，由其自生自灭，中国古典园林将如同传统绘画及其他传统艺术，逐渐沦为考古遗迹。诸多精美园林，若不及时采取措施，即将走向湮灭之境。人民政府已付巨大努力，提振世人欣赏与评价传统园林技艺之热情。吾人既已看到并认知这一危境，古老文明之花正在枯萎并迅速凋谢，后人以及整个世界，则有责任针对这些值得保护、值得欣赏之园林，给予中肯客观之评价。

◇ 网师园。

◇ 宁波，天一阁。

七　东西方比较 | 99

◇ 怡园。中国园林中，建筑如此赏心悦目，鲜活成趣，令人轻松愉悦，即便无有花卉树木，依然成为园林。

◇ 拙政园，塔影亭。建筑而非植物，主导景观，成观者视觉之焦点。（左页图）

七 东西方比较 | 101

◇ 狮子林。眼前世界，逐次展开，诗句铭文，激发想象，无边景色，引人遐思。

沿革 (八)

历史 （一）

今日所见中国园林，其发展雏形可溯至公元前约一八〇〇年。夏朝桀帝建玉台以为乐。《诗经》有载，六百余年后，周朝第一位统治者文王曾建灵台、灵沼、灵囿。该书尚提及果园、菜圃及竹林。

一座园林之构想，无疑始于游牧民族转入更为稳定之农耕文明。秦代始皇帝统治时期（公元前二四六至公元前二一〇年），皇家园林首次大规模兴建，"台地"、"池沼"与森林、动物园融合成为游猎场所。是园称作"上林"（皇家丛林），后被汉朝武帝（公元前一四一至公元前八七年在位）大加扩展，并与其他花园、别墅一起装扮首都长安，即现今陕西省省会西安。一位皇帝有限之生命时光，不幸与普通百姓同样短暂，致使武帝苦恼不堪。遂听从巫师，建造环水之人工山丘，以象征天堂乐土，冀助其长生不老。根据这一主旨，中国园林之营造，开始萌发池岛构想。后经移植，该构想也出现于日本园林中。

此时之园林，已不再限于天子独有之游乐场所。武帝之叔梁孝王，其华丽别墅地处今日河南省境内，以一座叠石假山而称著，记载证明，叠石为其首创。几乎同一时期，富绅袁广汉，纵情声色，趣味相投。其园林临近洛阳，亦处河南。武帝时期，出现董仲舒[1]之最早文人园。据传，仲舒为集中思想，尝垂帘三载，以免园林景色分其心智。彼时文人园林适可而止，仅于庭内种植若干树木，三两花卉而已。毋庸置疑，与欧洲中世纪之修道庭园颇为相似。

晋代（二六五至四二〇年），私家园林始兴。石崇既为船业巨贾，亦为文人墨士，其金谷园建于河阳，亦处河南。石崇挥霍无度，其嗜欲之一，令美人步于芦粉之上，有留浅痕者，即被视为过胖而须节食。宠姬绿珠，忠义如梅萨利纳[2]，以身殉情。

石崇自述，其园"金谷"，由树、湖、亭、楼阁组成，飞鸟游鱼，簇汇群集。石崇终日纵情声色，垂钓、赏乐、读书。稍后半世纪许，有顾辟疆者，既具公卿风度，又兼文人格调，于苏州拥有一座园林，为中国东部首次记载。

---

1. 董仲舒（公元前一七九至公元前一〇四年），西汉思想家、哲学家、政治家、教育家，广川（今属河北）人。曾建议汉武帝罢黜百家，独尊儒术，为汉武帝采纳，使儒学成为中国社会正统思想，影响长达两千多年。董仲舒曾提出天人感应、三纲五常等重要儒家理论。
2. 瓦莱里娅·梅萨利利纳(Valeria Messalina，约二二至四八年)，罗马皇帝克劳狄的第三个妻子，以淫乱和阴险出名。绿珠(?至三〇〇年)，西晋石崇爱妾，善吹笛，赵王伦专权时，赵党孙秀曾指名向崇索取，为崇所拒，后崇被逮，绿珠坠楼自尽。

辟疆享誉后世，直至唐朝，广受推崇，其园林同样广受乡士与游客艳羡。某日，辟疆曾与一位同样傲慢之擅入者恶辩，此不速之客正是王羲之后嗣[1]，父子二人皆为大书法家。

盛唐时期，都城长安，京畿近郭，别墅广布，园池群集。政客权贵，文人雅士，往来川流，消夏避暑，营园成风。其中最著者为诗人兼画家王维之辋川别业。是园占地广阔，山脉延绵，泉石占胜，景色不绝。如前所述，王维为其辋川别业绘制二十幅图景，佳山妙水，层出不穷，为后之雅士虔诚仿效，园遂美誉日增。另一唐代诗人白居易，无论身居何地，即便短驻，皆营园。其作虽未精巧，然观一丘一溪，以感自然，亦足以令其心满意得。

1. 指王献之。

北宋（九六〇至一一二七年）洛阳，私园甚多。李格非作《洛阳名园记》，所载园林逾二十五座，其描绘鲜动，令人赞赏。格非尤其强调营园之各类禁忌，以为过大水面、过多人工，皆可引发不利因素。时过境迁，"风水轮回"，皆可颠覆园林之命运，今朝此在，明日他去，着实令人为之忧虑。京都汴梁，亦因园著，皇家石园"艮岳"，名重一时。

南宋（一一二七至一二七九年）时期，杭州与湖州皆为园林之城。杭州亦为当时都城，除众多私家园墅，尚有皇家御园。著名西湖拥无限美景，周围山色使之倍增魅力。湖州为今日之吴兴[1]，山池荟萃，有园不下三十，现已荡然无存。今日二城，劣园填塞，全不复昔日光彩。

北宋另一园林城市苏州，有二园尤为著名。艮岳督石官朱勔[2]，乃一地方官吏，尽搜奇石，以尽其责，狠虐苛严，流威东南。朱勔于苏为己营墅，殚精竭虑，耗尽才华，取名"绿水"，今已不存。另一园林，朱伯原之"乐圃"，虽几经易主，难以辨识，但至今犹在，为"环秀山庄"。

时至明代（一三六八至一六四四年），私家园林已臻全盛，主要分布于江浙省域。最重要者，则在苏州。明代园林，虽饱经沧桑，至今仍幸存者，不在少数。时造园营圃，蔚然风行，超越玄知，而成一类专业技能。一六三四年，计成[3]所著之《园冶》，实为造园学之肇始。在此专著中，计成详述造园理论与实践，描绘各类风景之艺术，并辅以有关装饰之各种章节与图例。计成未以一章专论花卉树木，仅在各处相应内容中，简单述及。同期另一人物

---

1. 吴兴今称湖州市。
2. 朱勔（一〇七五至一一二六年），苏州人，北宋大臣，为"六贼"之一。因父亲朱冲谄事蔡京、童贯，父子都任有官职。因宋徽宗垂意于奇花异石，朱勔奉迎上意，搜求浙中珍奇花石进献，并逐年增加。政和年间，在苏州设应奉局，靡费官钱，百计求索，勒取花石，用船从淮河、汴河运入京城，号称"花石纲"。朱勔在奉迎皇帝的同时，又千方百计，巧取豪夺，广蓄私产，生活糜烂。钦宗即位，将他削官放归田里，后又流放到循州关押，最终斩首处死。
3. 计成，生于万历十年（一五八二年），卒年不详，字无否，号否道人，原籍松陵（今江苏省苏州市吴江区同里镇）。明代著名造园家，崇祯七年（一六三四年）完成造园著作《园冶》。

朱舜水[1]，于明清更迭之际，逃难日本，使日本园林深受中国之有力影响，尤其文人园林。今日于东京"后乐园"，仍可目睹其伟大成就。

清朝（一六一六至一九一一年）初期，扬州改观，成为史上最为辉煌之园林城市。康乾两帝，皆数次南巡，均尝造访。乾隆皇帝，既为艺术家又兼鉴赏家，游赏园林，喜题善咏，以诗叙之，并带走精选之山石佳品。瘦西湖实为悠长且宽狭不一之水道，沿之皆佳山水，北垞南陔，绵延不绝，自市肆至山林，俨然一道连续之绿链，构造无可比拟之城市意象。彼时扬州，六街三市，巨贾富商；舳舻相继，盐运全国；货殖通财，往来四方，除运回现钞外，亦有作为压舱之大块山石，随处可得，叠石之价得以平抑。

时花匠园丁，大量涌现，其中盛名最广、口碑最佳者，为王氏家族也。其园艺之术，精而有道，后裔至今仍以此为生。宽裕之财，易得之料，扬州富裕殷实之家，巨墅微庭之院，皆可掇山理水，营园造景。

扬州多数名园，唯一要旨即在取悦天子。为待秋季天子幸临，不吝辛劳，以足其愿。某日，乾隆巡游运河，

1. 朱舜水（一六〇〇至一六八二年），即朱之瑜，字楚屿，又作鲁屿，号舜水，浙江绍兴余姚人，明清之际的学者和教育家。清兵入关后，流亡在外参加抗清复明活动。南明亡后，东渡定居日本，在长崎、江户（今东京）授徒讲学，传播儒家思想，很受日本朝野人士推重。著有《朱舜水集》。

偶言某寺之旁，宜增白塔，以饰远景。噫乎！赞乎！次日，大自然因新添塔刹而大有改观，不消说，龙心悦然。然此塔构筑如此简略草率，以致不久之后，拆除重建。

乾隆在其数次南巡中，曾亲临无锡、苏州、杭州。所驻跸之园林，皆因临幸而山泽增辉。其中有遵其意旨者，于北京进行复制。除古罗马哈德良大帝[1]外，世间鲜有统治者，在其领地，广泛游历。因其所爱，遵常去之所，仿制复建，逾多亦善，以保持愉快记忆。乾隆逝后，扬州几乎所有园林都遭弃置、衰败，乃至最终消失，存者仅略二三。除众多私家园林外，著名者尚有寺庙园林小金山，以及平山堂。

南京迄今依然保存两座十四世纪以来之皇家园林，煦园与瞻园。因其不时为政府机关所用，得以留存，并获适时修护。

清初南京，最具盛名者，随园是也。园主袁枚[2]，既为文者诗人，亦为美食大家。一七四八年，袁枚于小山斜坡上，购得一处久已坍圮之别墅[3]，穷

1. 哈德良大帝（Publius Aelius Traianus Hadrianus，七六至一三八年），军事家，政治家，罗马帝国安敦尼王朝的第三位皇帝，五贤帝之一，一一七至一三八年在位。精通星象学，博学多才，爱好广泛，又具艺术家气质。他曾经重建毁于大火的罗马万神庙，一生游历甚多。
2. 袁枚（一七一六至一七九八年），字子才，号简斋，晚年自号仓山居士。钱塘（今浙江杭州）人，祖籍浙江慈溪。清朝乾嘉时期代表诗人、散文家、文学评论家和美食家。主要传世的著作有《小仓山房文集》《随园诗话》。
3. 袁枚于《随园记》中写道："康熙时，织造隋公，当山之北巅，构堂皇，缭垣牖……号曰随园，因其姓也。后三十余年，余宰江宁，园倾且颓弛，其室为酒肆……问其值，曰三百金，购以月俸。"
4. 普林尼（Pliny the Younger，约六一至约一一三年），古罗马作家、行政官。死后留下一批富有文学魅力的私人信札，这些信札描述了罗马帝国全盛时期的社会生活和私人生活。他出身官贵家庭，是作家老普林尼的养子，也被称为"小普林尼"。
小普林尼在两封信里留下关于古罗马园林最详尽的记录。他自己拥有两所别墅，一所在奥斯提亚东南十千米处，名为劳伦提安（Laurentian）；另一所在塔斯干（Tuscan）。
小普林尼非常得意于塔斯干别墅。他写道："除了所有上述赏心悦目之物外，还要加上一条，就是我在这里能够彻底放松，享受安静而隐潜的消遣……气候宜人，天空明朗，气息清新，一切都有益于身体和心灵。我在这里读书下棋，陶冶性情。"

其余生，培植此著名园林，恰如罗马之普林尼[4]（注释内容见左页）。袁枚为极少数幸运之文人，拥有私园，却供公众游乐。袁枚逝后，园林即遭弃置，偶有修缮。约存至一八五三年，咸丰兵劫期间，尽遭损毁。

清朝洪杨之役，毁园无数，旧迹凋零。然于此前此后，论质论量，江浙园林均为全国之冠，下节所述现存园林，皆属江浙两省[1]。

1. 其中一些园林今属上海市。

(二) 现况

宋代、明季乃至清初之园林，多所圮毁，旧观不复，仅存遗痕。兵劫，火灾，以及非耐久之材致使建筑脆弱易腐，名园佳墅，渐就凋敝，咸同兵火，致以最后一击。今日所见园林，余痕虽存，风韵不再，一蹶而难以复兴。现代材料，再添伤毁。旧式园林建筑某些极富魅力之特征，如花格纸窗，卵石路面，不复使用；玻璃窗格，水泥步道，取而代之。时代进步，亦为悲剧。商贾精神，艺术式微，园宅格局，屈于现代发展。中国传统园林，诚为人类最精美成就之一。如此境遇，令人痛心。

◇ 留园,石林小院。中国园林之营造,开始萌发池岛构想。

八 沿革 | 113

江苏
——
苏州

拙政园

拙政园，或曰满洲园，位于城之东北。十六世纪初，王氏献臣建于元代寺庙遗址上。不久因偿还赌资，归于徐氏。清初为陈氏所得。清朝驻军，扩充而改为将军府。一六七九年，为常州道署所在，旋即复为私有。约一七四二年，蒋氏整修复之，起名"复园"。此后一百年间，两易其主，咸丰兵燹时，尝为洪杨将领据为治事之所。兵乱之后，用作巡抚官邸。一八七二年，改为"八旗奉直会馆"，因有满洲园之名。一九一一年辛亥革命后，清廷撤离，园林遂对外开放。

园西乃另一园林，原系满洲园一隅，以墙分隔，后归叶氏，一八七七年为张氏所有，直至一九四九年。（拙政园或满洲园，意为"拙者之园，不谙政务，唯愿种艺"。）

◇ 拙政园，见山楼。

◇ 拙政园，补园折廊。

◇ 拙政园，倚玉轩。

◇ 拙政园，小飞虹雪景。

江苏

苏州

狮子林

狮子林，邻近拙政园，原为宋代官吏产业，至元代售予寺院。主持寺僧惟则[1]，一三四二年始营叠石，经年累积，遂成今日所见之假山。石山声名远播，影响广泛，部分缘于画家倪瓒[2]，园成四十年后，倪瓒作《狮子林图》。明末之际，园寺分开。乾隆皇帝曾造访此园，时属黄氏。高宗深爱此园，乃至在热河行宫仿制相同格局之园林，后在北京又造另一。狮子林之称，似有两种解释，其一在于石山，一峰叠石，正似此兽；其二，另有人认为，惟则来自浙江天目山邻近所谓"狮子峰"之寺院，因而将其在苏州叠石命名为"狮"，以志旧时之忆。一九一七年，该园转予贝氏，贝氏以染料交易而致百万巨富。园内假山无大变动，其余部分主要于二十世纪二十年代完成。画梁雕栋，鬼斧神工，以炫主人之富有。

---

1. 惟则（一二九八至一三六八年），元僧人、文学家。又作维则，字天如，本姓谭，名吉之，庐陵（今江西吉安）人。髫年出家，初参海印，后得法余杭吴山高僧明本（号中峰）。至正二年（一三四二年），从浙江来吴（今苏州），其弟子相率出资，买宋代废园建禅林，供其起居，名狮子林，成苏州名园。
2. 倪瓒（一三〇一至一三七四年），元末明初画家、诗人，号云林子。江苏无锡人。倪瓒家富，博学好古。倪瓒擅画山水、墨竹，师法董源，受赵孟頫影响。早年画风清润，晚年变法，平淡天真。疏林坡岸，幽秀旷逸，笔简意远，惜墨如金。以侧锋干笔作皴，名为"折带皴"。墨竹偃仰有姿，寥寥数笔，逸气横生。存世作品有《渔庄秋霁图》《六君子图》《容膝斋图》等。

◇ 狮子林，雪景。

◇ 狮子林，真趣亭。

◇ 狮子林，中部水景。

江苏 — 苏州

留园

留园，原名刘园，位于城外，最早可溯至十六世纪中叶，为徐氏东园。一七九四年，归刘氏，遂名。一八七六年为盛氏所得，新主希望保留园林旧有声誉，维持其音不变，重新命名为留园，以寓"停留片刻"。盛氏既富且贵，增拓新建，花费无度。留园中部极为精彩，东部则为一座错综非凡之迷宫，由为数众多之庭院与游廊构成，其间包容精选奇石。总体格局，宏丽轩举，胜境天成。

◇ 留园，小蓬莱。

◇ 留园，涵碧山房。

江苏
——
苏州

◇ 环秀山庄，假山。

环秀山庄

环秀山庄，景德路旁，原为五代（十世纪）一座皇家园林。宋属朱伯原，名"乐圃"。元季归张氏。约一四七〇年，售予杜东原[1]，后又转归仕人申时行[2]。清乾隆年间，经蒋氏手。假山石洞，小而精巧，远胜狮子林之巨堆，并可用作密斯[3]所谓"少即是多"之完美例证。咸丰兵劫，略有损坏。一八九八年修复。二十世纪三十年代为汪氏所得，其时赠予家族，作为福利机构，从此亦称"汪义庄"。

---

1. 杜东原（一三九六至一四七四年），即杜琼，字用嘉，号东原耕者、鹿冠道人，明朝南直隶苏州府吴县（今属江苏）人。明经博学，旁及翰墨书画皆精。山水宗董源，层峦秀拔，亦工人物。好为诗，其诗于评画尤深。著《东原集》《耕馀杂录》。
2. 申时行（一五三五至一六一四年），字汝默，号瑶泉，晚号休休居士。明代大臣。明朝南直隶苏州府（今属江苏）人。嘉靖四十一年（一五六二年）殿试第一名，获状元。历任翰林院修撰、礼部右侍郎、吏部右侍郎兼东阁大学士、首辅、太子太师、中极殿大学士。
3. 路德维希·密斯·凡·德·罗（Ludwig Mies van der Rohe，一八八六至一九六九年），德国建筑师，与赖特、勒·柯布西耶、格罗皮乌斯并称四大现代建筑大师。密斯提出"少即是多"的建筑设计哲学，在处理手法上主张流动空间的新概念。

◇ 环秀山庄,问泉亭。
◇ 环秀山庄,假山。(右页图)

江苏
——
苏州

怡园

怡园，自得之园。位于城市中心，明代吴姓住宅之西。现存园林主要部分为十九世纪九十年代，一位富有退休官僚顾文彬[1]所筑，其后裔（大多为文人与画家）加以扩充，并因此在布局中留有诸多影响，但效果不甚理想。其因在于，苏州拥有太多佳例可资仿效，反成障碍，不利于做出最优选择。

园内收藏宋代、明代家具，嘉木莳花，数不胜数，称著一时。

◇ 怡园，秋景。

---

1. 顾文彬（一八一一至一八八九年），字蔚如，号子山，晚号艮盦，元和（今江苏苏州）人。道光二十一年（一八四一年）进士，官浙江宁绍道台。

◇怡园,画舫斋。

◇ 怡园，画舫斋。

八 沿革 | 135

江苏
苏州

沧浪亭

沧浪亭，号称苏州现存最古之园林。始于十世纪，节度使孙承祐之别业。约百年后，宋代文人苏舜钦[1]购置其处，于水边起亭，遂成著名之沧浪亭，亭名亦成该园之名。十四世纪，宅园转为僧院。十六世纪中叶，僧人文瑛复制一亭，沿用前名，僧院又转回园林。一六九五年，园地扩增，沧浪亭移至小山顶部，即现在位置。一八二七年，园林修缮，零星增添若干新屋。咸同兵火，园林受损。一八七三年，进行大量重建。小坂逶迤，花木扶疏，颇具郊野之气息，是园与苏城及他处众多园林有所区别。入口大门之东，有厅，名"面水轩"，不带片尺园墙，园林与外邻池水彼此衬映，互为交融，景色拓展。该法实为独创，别处少见。园墙附带颇多漏窗，曾有一百零八处，却无两者相互雷同。嵌花露孔，式样变幻，逸趣横生，亦为佳话。

---

1. 苏舜钦（一〇〇八至一〇四八年），北宋词人，字子美，祖籍梓州铜山（今四川中江），曾祖时迁至开封（今属河南）。因支持范仲淹的庆历革新，为守旧派所恨，御史中丞王拱辰让其属官劾奏苏舜钦，罢职闲居苏州。后来复起为湖州长史。著有《苏学士文集》诗文集，《苏舜钦集》十六卷。

◇ 沧浪亭。

◇ 沧浪亭，葑溪。

江苏 — 苏州

网师园

网师园，始于一一七四年，退休官僚史正志[1]之别墅，称"渔隐"。几世纪来，池馆萧条，渐就荒废。约一七六五年，园产归宋宗元[2]，重修复建，名"网师"，以复古意。此后，该园再遭弃置。十八世纪晚期，瞿远村购之，复修增建，改称"瞿园"，并于布局方面建树颇多，遂如今日所见。[3]后售予吴氏。洪杨兵火后不久，约一八七〇年，再度转让退休官僚李鸿裔[4]。李于斯地，居之较久。其后裔又将园售予清廷博学将军达桂[5]。"中华民国"成立之前，除达桂家族外，曾另有清代官员兼文人张金坡[6]购得此园，但从未观其一眼。其戚何亚农[7]于二十世纪三十年代购得此园，并保存至五十年代，后归国家所有。

寓居网师之画家，先后不止一位。园内美景与浪漫气息，无疑有利于艺术创作，激发灵感。

园林中央，水池一方，略有建筑，将景色分为南北两区。南部为主人娱乐之所，北部则为读书之处，以适更为正

---

1. 史正志（一一一九至一一七九年），字致道，丹阳人，南宋吏部侍郎。
2. 宋宗元，字少光，清江苏元和（今苏州）人。乾隆三年（一七三八年）中举，任天津道。乾隆三十六年（一七七一年），运河涨溢，天津城西茶园堤溃，洪水泛滥南乡，复逆行围城。宗元闻舆人言，向直隶总督献泄洪之策，惜其议未得付诸施行。迁光禄寺少卿，后辞官归家。
3. 乾隆末年（一七九五年）太仓富商瞿远村购网师园，增建亭宇，叠石种树，半易网师旧观，有梅花铁石山房、小山丛桂轩、月到风来亭、竹外一枝轩、云冈诸胜。至今该园尚总体保持着瞿氏当年造园的布局。
4. 李鸿裔（一八三一至一八八五年），字眉生，号香严，又号苏邻，四川中江人。咸丰元年（一八五一年）举人，官至江苏按察使加布政使衔，官兵部主事。罢官后，家苏州。
5. 达桂，光绪三十一年（一九〇五年）至光绪三十三年（一九〇七年）任吉林巡抚。
6. 张金坡（一八四三至一九二二年），名锡銮，浙江钱塘（今杭州）人，生于成都。一九一七年，张作霖以三十万银圆在苏州购得网师园，赠予其师张锡銮庆寿。
7. 何亚农（一八八〇至一九四六年），原名何澄，字亚农，别号真山，山西省灵石县两渡村人，著名文物鉴赏收藏家。一九〇八年于日本陆军士官学校毕业，早年加入同盟会，是国民党元老之一，后退出军政界隐居苏州经营实业及教育。

规事宜。园之东部有连续庭院，与厅堂相续，足纳大型家庭。园内建筑极为精湛，林木配选雅致，布局优美。承蒙布鲁克·阿斯特夫人之捐赠，一九七九年，网师西北角庭院与建筑，因其恬静之特征，选为中国园林之样板，由来自苏州之一组匠人，于纽约大都会艺术博物馆某一采光庭院内，得以复建，永久展览。

阿斯特庭院所纳之苏州园林，于材料与工艺方面无可挑剔，然本质问题犹存：园林这般充满生气之活物，能否按照无生命之博物馆珍品，摆放陈设，供人观览？一座中国园林，或其局部，能否移植大洋彼岸，进行东西方之合成？采用砖瓦复制园林是一回事，再生复兴，激发生命精神，则全然另一回事。一篇刊于《进步建筑》(*Progressive Architecture*, 8/1981)上之精妙评论，值得引用：

"二者永远不能混为一谈……毋庸置疑，整个过程如此严谨，结果却令人失望至极。行走于阿斯特庭院，与真正中国园林所获之体验，鲜有相似之处。其细节严格拘谨。穿过月门，进入明轩，沿蜿蜒之瓦顶游廊，信步向前；中国格调之青枝绿叶，透于两侧门洞漏窗。岩峦劈石，奇渍斑斑，泉水潺潺，萦回绕流；墙垣被完美精细之檐口所平分，仿佛界墙，但其实不然。因明轩置于室内，建于光井。檐口之上，墙仍延续，直抵人字形天窗。近墙处，瓦铺屋面茫然消失，漏窗后之竹林，种植于浅薄壁龛，并辅以人工照明。这并非无足轻重之转换问题。中国悠久精深之传统内涵，体现于每座园林，并浸透至每一要素，从而造就一种无限延伸之景观。能工巧匠，珍材妙手，此处却被四下之隔墙，密布之直线所彻底征服。因为在此，天窗之下，大都会艺术博物馆所展示，是一所静谧庭院，精致展品与复制片段，悬挂并填嵌于其中，恰如其他西式博物馆所习以为常之方法(N. M.)。"

较"大都会"更甚一步,巴黎之乔治·蓬皮杜国家艺术与文化中心,将同一苏州园林按1∶50比例做成模型,拟于一九八二年某时在中心展出。展场提供鸟瞰图景,并通过电子方式复现园林各个部分。此法似有效乎?谨请牢记,歌德曾云:"仅当真正绕行并游走其中时,建筑生命才能得以体验。"歌德之论,必可用于园林,因其比一幢房屋更具生命力,更为错综复杂。充分欣赏一座真实而非模仿中国之园林生活,必须亲历其境,漫步逗留,立坐进出,认真体验整座园林院墙内之每一角落。

◇ 网师园。

◇ 网师园。

◇ 网师园。

江苏
——
苏州

西园

西园，距留园不远，一度曾为其附园。十六世纪，徐氏家族持有东园（今留园），故命名此分离邻园为"西园"。后赠予僧院，成寺庙园林。咸同兵燹，佛殿园林，俱成灰烬。旋即于一八九二年，又得重建。

园林中央，碧水一池，其中又坐落巨亭一座。寺院塘池用于佛教善行，将钓者在别处捕获之活鱼，在此放生。长久以来，园林已与庙宇隔离。

◇ 西园。

八 沿革 | 147

江苏 —— 苏州

耦园

◇ 耦园。

耦园，双生之园，其现有场地东部之起源，最初应属清初退休官吏陆锦[1]之涉园。其后几经易主，最后于一八七五年，为退休官僚沈秉成[2]购得。旧园不仅全部修复，进而合并西部院落，于此另建一较小园林，整体园墅为之一扩。园中，存有一座珍本藏书楼，并以湖石装饰庭院。东西两园并置，主人因而称此联合为"耦"。民国时期，全部园产归纺织实业家刘氏，直至一九五〇年，刘氏赠予地方政府。

两园规模并不均等。东部较大且制作更为精致，辅以令人称羡之黄石假山，园之大部于清初之际即已建成。东园假山充满自然魅力，可为样板。苏城尚无其他同类可以与之媲美。

---

1. 陆锦，清初顺治年间保宁知府，原籍苏州，雍正年间致仕归里。
2. 沈秉成（一八二三至一八九五年），清官员、藏书家。字仲复，自号耦园主人，浙江归安（今湖州市）人。同治十三年（一八七四年），因病寓苏，携妻归隐，购得涉园废址。曾创办南京水师学堂、经古书院等教育机构，著有《蚕桑辑要》。

江苏 —— 苏州

艺圃

艺圃，又名药圃，十七世纪初，为著名文人，画家文征明[1]之曾孙文震孟[2]所建。十七世纪中叶，改名"艺圃"（艺者之园）。

池水位于园之中央，布局得当，将这中等规模之园林，隔为南侧假山园林与北侧一组建筑。假山密布繁茂植物，精巧石峰，蜿蜒曲径，均为优秀样板。但其朝向有误，导致自北岸观赏而遇眩目逆光，效果因此减弱。

水池西南，有小院，以墙分隔，中有小溪、叠石，与建筑物相接，异常精美。"乳鱼亭"位于池之东岸，其木构架可溯至明代。

---

1. 文征明（一四七〇至一五五九年），原名壁，号衡山居士，世称"文衡山"，长洲（今江苏苏州）人，明代画家、书法家、文学家。因官至翰林待诏，私谥贞献先生。五十四岁因为岁贡生荐试吏部，授翰林待诏，任官不久便辞官归乡。留有《甫田集》。文征明诗、文、书、画，造诣极高，人称"四绝"全才。其与沈周共创"吴派"，在画史上与沈周、唐伯虎、仇英合称"明四家"或"吴门四家"。
2. 文震孟（一五七四至一六三六年），明代官员，书法家。长洲人，文征明曾孙。天启二年（一六二二年），状元及第，授翰林院修撰，崇祯初拜礼部左侍郎，兼东阁大学士。因疏陈勤政讲学，忤魏忠贤，被廷杖八十，贬职调外，愤而告归故里。

◇ 艺圃，浴鸥小院。

◇艺圃。

江苏——苏州

鹤园

鹤园，苏州众多小型园林之一，清朝末年之际，退休官僚洪鹭汀[1]修建。

入口门厅为五开间建筑，或许过于正式显赫，敞对园林，经由连串曲折游廊，访者得以穿行不同建筑之间。水池位于园林中央，周边环以叠石、建筑与植物。

三座几乎同样之厅堂成为园林主景，该园以两种手法极大缓解由此产生之单调。其一为邻近西墙之"梯形馆"，其二为带有六角凉亭之南侧小丘，园林四周均配有精选植物。

另有若干小园值得一访。离开苏州之前，还应游览拥翠山庄，为虎丘山坡一座台阶式园林。

◇ 鹤园。

---

1. 洪鹭汀，清朝道员。光绪三十三年（一九〇七年），洪鹭汀在宅西隙地筑园，工未竣而离苏。

八 沿革

江苏 — 苏州

拥翠山庄

拥翠山庄，迄今所知该类园林之罕见样板。一八八四年，朱氏及其友人营建。

约十一点五米高差内，一条狭长山地分为四层平台，其上置落房屋、林木与叠石。阶台之上，游者得以极目眺望，园林台景，四周郊野，融为一体，将视野延至地坪，虎丘之塔亦可尽收眼底。

◇ 拥翠山庄。

江苏 —— 苏州

高义园

高义园,为另一座山地园林,位于苏城之西,约十千米,地处著名景区天平山麓。晚秋时分,天平遍布红艳枫树、柑橘以及栗色栎树。有传闻言,尝有年轻文人,寒舍居焉,天平佳景,怡然载之。该文人并非他者,正为十一世纪北宋伟大政治家范仲淹[1]。园产为后嗣增扩,呈多层阶台之别墅,树丛、溪泉、巨砾和石洞,为其增辉。清帝乾隆,曾访此园,题"高义园"之称,为示其对范仲淹道德品质之敬仰。十一月时,为游赏最佳季节。

◇ 高义园。

---

1. 范仲淹(九八九至一〇五二年),字希文,苏州吴县人。北宋思想家、政治家、文学家。一〇一五年,范仲淹苦读及第,授广德军司理参军,后历任兴化县令、秘阁校理、陈州通判、苏州知州等职,因秉公直言而屡遭贬斥。

江苏
——
苏州

残粒园

残粒园,建于清末。原为盐商宅院之小庭园,仅一百平方米。一湾碧池,人工挖凿,湖石衬砌,若干凉亭环绕院墙,建于山顶,为小型园林建筑之典范。

◇ 残粒园。

江苏
——
吴江

退思园

退思园，"退而思过"，位于水乡同里，任兰生[1]所有。兰生因受弹劾罢官，退休返乡，一八八五年，于宋代别墅原址上，建此宅园，以示悔悟之心。兰生援请诗人画家袁龙[2]设计。为表明谦卑之觉，着重强调自律克己，因此，园师才华未能充分施展。

园之中心为一大厅堂，一池在前。不久以前，此园常为著名画家、诗人聚会之所，其中有长时寓居逗留者。

◇ 退思园。

---

1. 任兰生（一八三七至一八八八年），字畹香，江苏震泽（今同里）人。同治三年（一八六四年），捐升同知候选，后投安徽巡抚乔松年，三十岁晋升道员，不久加布政使衔，驻防寿州。光绪五年（一八七九年），官至凤（阳）颍（州）六（安）泗（州）兵备道，遭内阁学士周德润弹劾。光绪十年（一八八四年），任兰生四十七岁时被革职，回到老家同里。
2. 袁龙，同里画家，字东篱，诗文书画皆通。

江苏 — 扬州

◇ 何园，假山。（右页图）

何园

何园，雅称"寄啸山庄"，"美音传扬之别业"，十九世纪八十年代，退休盐税行政官何芷舫[1]建于一园林旧址。今为扬州最大且最著名之私家园林，其规模以及营造技艺，堪与苏州园林媲美。

中央庭院饰以水池、凉亭与山石，邻近背景为一座二层之宽敞厅堂。一道悠长曲折之游廊，隐约遮隐建筑中之各类活动。邻近一组房屋，曾为主人居所。园内精选植物，尤其古树嘉木，值得留意。

---

1. 光绪九年（一八八三年），何芷舫由湖北汉黄德道兼江汉关监督的官任上卸任，到扬州购得吴氏片石山房旧址，后扩为园林。

◇ 何园。

江苏
———
扬州

个园

个园，如此称谓，实为此园主导之竹林，以"个"为文字单元，形象化刻写"竹"字书法。十九世纪初，某一黄姓盐业巨商，开创园林肇始。黄氏以个园为笔名。为强化其本人与园林之关系，全园遍地植竹。另一独有特点即为"四季假山"。四座叠石小山，分布园林四周。石材皆精心挑选，色彩与种类、形状与朝向之安排，配以恰当植物分布，均显示设计者联想之才能：呈现自然山岭于各种特定季节中之感觉与景象。四座假山皆有石洞、天井、台阶，观者可于每一山顶上，远眺郊区。

◇ 个园。

◇个园。

江苏
——
扬州

棣园

棣园，清初即已存在。先后属陈氏、黄氏、洪氏，最后为包氏所有。十九世纪中叶，包氏命名此园为"棣园"，随后并入湖南会馆。全园图曾于一八四五年勒刻于石，原为墨绘草图摹本。

江苏
——
扬州

凫庄

凫庄，"野鸭别墅"，为溪流中一小岛，陈氏始建于一九二一年，坐落于一小岛上，并以一桥与岸相连。别墅可综览四面八方之优美风景全貌。三十年代陷入毁圮，今日经整体布置，才重新完成。

前述园林为扬州最著名者。小型与中等规模园林，不胜枚举，数量仅次苏州。目前多数园林正处于修理复原之中。

扬州历来即为富贾达官、盐业新贵汇居之地。因社交活动需要，园林拥有设备完善之宴会厅堂，露天戏台。这类特殊要求，使之不同于苏州以及其他各处园林。

江苏

泰州

乔园

◇ 乔园。

乔园，其源可溯至十六世纪，明代达官陈应芳[1]于"涉园"（漫步之园）原址建造。十七世纪中叶，园产归田氏。约六十年后再度易手。郭氏得之，命名为"三峰园"（三峰冠于山巅之园）。约十九世纪五十年代，一富有盐官购得此园，遂成最终主人。

园之中央景点，为五开间厅堂，南侧有一U形流泉，以叠石镶边。附近石洞乃明代所构。由于附近缺乏足够所需叠石，遂以湖石与黄石混用，或以砖为核，覆以散石砌面，以权宜之计，堆叠装饰性石山。

中央厅堂之正面为一独立庭院，主厅两侧，各辅以凉亭。园以浓荫古木、精美花卉而称著。

---

1. 陈应芳（一五三四至一六〇一年），字元振，明朝南直隶泰州卫（今江苏泰州）人。万历二年（一五七四年）进士。官至福建布政司参政，讲求探索水道原委与河之利害，悉其形势，辑录当时论河事资料，成《敬止集》。

江苏
——
如皋

水绘园

◇ 水绘园。

水绘园,明代文人冒辟疆[1]隐居之所。空间开阔,建筑精美,尝为明代诸多著名文人最爱常驻之所,随后衰落乃至塌圮。现园为乾隆年间王氏重建。拥有三进房屋,间以庭院。是园风格接近扬州园林。水绘园原名,自建成时起,一直沿用至今。

---

1. 冒辟疆(一六一一至一六九三年),即冒襄,号巢民,明末清初的文学家,明朝南直隶扬州府泰州如皋县(今江苏如皋)人。康熙三十二年卒,年八十三,私谥潜孝先生。冒襄一生著述颇丰,传世的有《先世前征录》《朴巢诗文集》《岕茶汇抄》《水绘园诗文集》《影梅庵忆语》《寒碧孤吟》和《六十年师友诗文同人集》等。其中《影梅庵忆语》回忆了他与董小宛缠绵悱恻的爱情。

江苏
——
南京

◇ 瞻园。

瞻园

瞻园，"仰望月宫"之园，始于十四世纪后期，以"西园"之称，毗邻徐达[1]之御赐宅第。徐曾为明代开国之功臣。园之精华位于东北，假山之下，藏有七窟，平台踞上，古木蔽焉。所有叠石，均采自著名的苏州太湖。紧邻假山前方，一不规则池塘，与南端另一较小水池以渠相连。巍峨厅堂，坐落其间。

明朝覆灭，瞻园纳入清朝卫戍总部衙门。乾隆皇帝曾访此园，赐名为"瞻"，以示愉悦之情，并下谕在北京颐和园东侧仿建。咸同兵燹期间，蒙受损坏，一八六七年旋即进行复原。一九〇四年、一九〇六年及一九四二年，均事修葺。临街入口，四进院落，精美假山，为一九六二年所建。

---

1. 徐达（一三三二至一三八五年），字天德，濠州钟离（今安徽凤阳）人，明朝开国军事统帅，淮西二十四将之一。

江苏

南京

煦园

煦园，"日照和暖"之园，始建时代与瞻园同，曾为明代开国皇帝亲属之园[1]，以长方水池为主题，近南岸处有石舫。向东不远，假山之巅建有套方凉亭，为园林建筑罕见。

明清易代，园林及其东侧宅第，为总督占据，省府所在，直至一九一一年辛亥革命。咸同兵燹时期（一八五一至一八六四年），曾遭中断。太平天国领袖占园为宫，故该园未遭损毁，自一九一一年后，用为公共机构。

◇煦园。

---

1. 指黔宁王沐英宅第。

上海

豫园

豫园，意为"舒适愉悦"之园，位于老城厢，临近城市佑护神庙（"城隍庙"），退休官吏潘允端[1]始建于一五六〇至一五七七年。南侧为内园，始于一七〇九年，称"东园"。一七六〇年，豫园归属城隍庙，取名"西园"，以与邻园有别。内园规模仅为总园八分之一。豫园之中，建筑面积宽广，设计精良，工艺高超，除了苏州外，似无可匹者。西北角黄石假山为造园巨匠张南阳[2]所叠，因而称著。张为十六世纪掇山专家。紧靠南侧边界一小溪旁，矗立一单峰名石，曰"玉玲珑"（灵巧雕琢之玉），为自然抽象雕塑之杰出范本。

内园，虽小，结构精巧，建筑、小丘、溪流布局雅致。因周边墙垣均充当建筑外墙，令人难以觉察园林边界。

---

1. 潘允端（一五二六至一六〇一年），字充庵，上海人，其父潘恩，字子仁，号笠江，官至明朝都察院左都御史和刑部尚书。
2. 张南阳（约一五一七至一五九六年），明代造园家。号小溪子，人称卧石山人。上海县（今属上海市）人。自幼从父学绘画，后以绘画构图造型法、叠造假山和造园而著名。上海县潘允端的豫园、陈所蕴的日涉园和太仓王世贞的弇园，都出自张南阳之手。

◇豫园。

上海
——
南翔

猗园

　　猗园，意为"茂盛精美"之园，建于十六世纪，原为闵氏园，后经李氏，命名为"猗园"。一七四八年，园归叶氏，善加修葺，更名为"古漪园"。一七八八年，园捐寺庙，后为私家所有。一八〇八年，复为另一家族所得，再臻完善。洪杨役后，园于一八六八年得以修复。一九二一年，此园完全圮毁，但又于二十世纪三十年代复兴。

　　一九八〇年，该园又事一轮大整修。针对园中建筑、种植、步道以及山水等方面之研究，渐次开展，力争成为全新完善之风景娱乐场所，以适应假日从上海涌来之游人。

◇ 猗园。

江苏
——
无锡

寄畅园

寄畅园,"寄情放怀"之园,仕人秦金[1]初建于十六世纪初,以惠山为园林衬景,以寺庙原址为园林基地。其后裔继承此业,曾孙易园名"寄畅"。清朝皇帝康熙与乾隆二人南巡时,曾先后访游此园。园以池底及地表众多流泉而著称。造园者充分利用水源,使之潺潺低语于假山之间,滔滔不绝涌于平台之下。明季之时,四周众多泉源,曾孕育多达八处久已消失之园林。

无锡城南,太湖之滨,曾有若干园林,二十世纪三十年代,多半为实业富商所有。

◇ 寄畅园。

---

1. 秦金(一四六七至一五四四年),明两京五部尚书,字国声,号凤山,常州府无锡县(今属江苏)人。成化二十二年(一四八六年)中举人,弘治六年(一四九三年)中进士。时称他为"两京五部尚书,九转三朝太保"。据秦氏宗谱所述,秦金为宋龙图阁大学士秦观之后。

江苏

——

常州

近园

近园，"亦为一园"，文人官吏杨兆鲁[1]初建于清初，园虽难称完美，兆鲁亦谦逊表示满足。池中有一座石窟假山，主厅位于其后。园林周边其余建筑皆以游廊相连。

一六七〇年，众多著名文人与画家曾为此园常客。园约于一八七〇年售予刘氏。数年后，为最后主人董氏所购。

---

1. 杨兆鲁，顺治九年（一六五二年）进士，历官至福建延平道按察副使。

江苏
——
昆山

半茧园

半茧园,曾为叶氏园,初建于一五四六年,后经数代扩建,最后命名为"茧"。清代拆分,叶氏后裔占其半,曰"半茧"。稍后此园为陆氏所购。十八世纪中叶,园林捐予寺庙,增添几座新建筑,四周环以墙垣。一八〇一年与一八二三年,再次进行改建与修葺,在此期间,一峰"寒翠石"自邻村移至园内山顶。此石为抽象雕塑之样式,最早可溯至十一世纪。

上海
——
松江

醉白池

醉白池，"陶醉诗人"之园。长期以来，松江即为府城，上海属之。清代此城富于园林，随后相继消失。醉白池仅为遗留下来之有趣物。清康熙时（一六六二至一七二二年），此园为一顾姓画家所有。十九世纪初，属于某基金创办之救济院。

◇ 醉白池。

上海
——
青浦

曲水园

曲水园,"蜿蜒小溪"之园,一七四五年为韩氏所建,为寺庙园林。十九世纪初,始名"曲水",因其二十四景而著称。一八六〇年被毁,十九世纪后期得以复原。

◇ 曲水园。

上海
——
嘉定

秋霞圃

秋霞圃,"秋季落日",始于十六世纪初,为高官龚弘[1]所建。十七世纪中叶,园属新贵汪氏。汪氏竭力维持园林幽雅,增添其魅力。一七二六年,园林并入邑庙。约四十年后,是园与申氏东园合并。咸同兵燹,受损极大,一八八六年曾进行紧急修葺。

◇ 秋霞圃。

---

1. 龚弘(一四五〇至一五二六年),字元之,一字蒲川,明南直隶苏州府嘉定(今属上海)人。成化十四年(一四七八年)进士,授严州推官。

江苏
—
江阴

适园

适园,"恬静舒适"之园,属清代翰林文人陈氏,建于一八五四至一八六四年间。程氏既为诗人也是画家,选场地之半造黄石假山。山之南为水池,岸边面山为主厅,以廊与东侧主人画室相连,另有一厅在山后北侧。此园规模适中,拥有种类丰富之建筑与植物,为其他园林所少见。二十世纪七十年代进行全面复原,不免丧失些许魅力。

◇适园。

江苏 —— 太仓

亦园

二十世纪三十年代，太仓仅有破旧园林二三处。蒋氏所属"亦园"（"亦为园林"）之一部分，即便部分已被征为医院，依然维持良好状态，具有做工精巧之建筑。所有园林于"二战"期间迅急消失。今日之地方政府，励精图治，以复旧观。

江苏 —— 常熟

丁氏园

丁氏园，晚清科举候补举人丁祖荫[1]所建。园依宅建，一掌之地。小池仅四至五平方米，与园东潺潺小溪相连。园内安置湖石若干，园以曲折廊道接于住宅。此园为现存少数私家小园之一。

---

1. 可能为丁祖荫（一八七一至一九三〇年），原名祖德，字芝孙、之孙，号初我、初园居士，又号一行。江苏常熟人。少年就读江阴南菁书院，清光绪十五年（一八八九年）庠生。中国近代知名官吏、学者、藏书家、文学家。

江苏 —— 常熟

燕谷

燕谷，"燕之峡谷"，一七八〇年退职于台湾知府之蒋元枢[1]所建。五十年后，园归同族蒋因培[2]，蒋因培遂委托叠石名师戈裕良构筑水岸山居，曰燕谷，以提示自己，逃离宦海之苦，享受归乡之乐，有如劳燕归巢。一八四七年园售予归氏。十九世纪末，蒋氏再度购回，但旋即为翰林文人张鸿[3]所有，张氏为最后主人，并进行大量修复工作。该园林木精致，匠艺高超，实为中国东南最佳园作之一。

◇燕谷。

---

1. 蒋元枢（一七三八至一七八一年），字仲升，号香岩，江苏常熟人。历泉州厦门同知，身兼台湾道，建多处炮台、书院与灯塔，并编修《台郡各建筑图说》。
2. 蒋因培（一七六八至一八三九年），清诗人。字伯生，常熟人。年十七以国子监生应顺天乡试，为法式善激赏。
3. 张鸿（一八六七至一九四一年），民国藏书家。初名张澄，字师曾，一字诵堂，别署隐南、橘隐、晚号蛮公，又称燕谷老人。

江苏——常熟

虚廊居

虚廊居，"茫然之园"，约一八九四年，退休文官曾之撰[1]建于一明代园林之荒废故址上。是园水景为主，成为此园引入入胜之缘由，建造茶亭，以便盛夏之际，迎客待人。

◇ 虚廊居。

---

1. 曾之撰（一八四二至一八九七年），字圣舆，一字铨仲，号君表。江苏常熟人，同治光绪间刑部郎中。品性洒脱，负有文名，交游遍公卿，与文廷式、张謇、王懿荣并称"四大公车"。辞官回常熟后在明小辋川遗址上筑"虚廊居"。著有《虚廊居诗集》六卷，《杂著》二卷，《群玉楼读碑记》二卷，《碑目》四卷，《江苏金石目》四卷。

浙江

嘉兴

浙江省因其传统园林而与江苏省齐名，即使数量略少。杭州为最著名之休闲城市，大部（虽非全部）别墅与园林已被现代化洪潮席卷而去，仅余郭庄。

嘉兴遗存中，有南湖烟雨楼，其源可溯至十世纪。是楼可瞰美妙景色，兼作茶室。城北有落帆亭，为邻近古运河河闸之小园，始建于宋代。

直至二十世纪三十年代末，南浔乃为虽小尤富之镇，并以园林而著称。如庞氏"宜园"，张氏"适园"，刘氏"小莲庄"。小莲庄后因主人收集书籍而闻名遐迩，园中有藏书阁。

◇ 烟雨楼。

◇ 落帆亭。
◇ 小莲庄。（右页图）

浙江 —— 海盐

绮园

◇ 绮园。

绮园，"绮丽之园"。清初建于更早宅园之旧址上。十九世纪属于一位在仕文人黄氏，稍加修葺后，命名为"拙宜园"（"适愚者园"）。咸丰时毁于兵火，其婿冯氏，承此地面仅存一古树之荒地，随即开始山池规划，并名之为"绮园"。由于缺乏钱款与文人情趣，此园魅力稍逊。仅存两处不同寻常之精巧设计：悬架于二垂直壁峰顶间之石梁，宛如飞桥，水中一道模仿杭州西湖之长堤。不久以前，此园正处修葺中。

浙江
——
平湖

莫氏园

莫氏园。莫姓家族,源自福建。清初,以经营木材起家。莫放梅[1],为莫氏成员,定居上海,仍与平湖多有往来。购得一片空地,始建住宅与园林,以强化家族纽带。

是园包含两块长方场地,用作采光庭院。院内有植物、盆景及若干室外家具。此二院与诸多迷阵式散布房屋相邻接,其中几幢为两层楼。

◇ 莫氏园。

---

1. 莫放梅(一八二七至一八八九年),长于画兰,博收名人书画,捐官正五品江苏直隶州知州。

八 沿革 | 211

◇ 莫氏园。

此前所列数十处苏州园林，为中国园林艺术最佳样板之荟萃。因地处温带，气候宜和，园林建筑往往筑以轻型构架，缀以精致细部，一园之各类植物，亦繁茂生长于其他园中。叠石假山，荷塘莲池，临水凉亭，各处相同。多数园林，若非文人、诗者、画家所有，则为其常去之所。谨此提示，如前所述，苏州园林之精美与数量，冠绝华夏。

Glimpses of Gardens in Eastern China

Foreword

Glimpses of Gardens in Eastern China

Talking about gardens in Eastern China, one invariably refers to those in Soochow (苏州) as the acme of traditional Chinese landscape art, yet a Soochow garden is not different from any other garden in this part of China so far as typical and common characteristics are concerned, and hence see a Soochow garden, you'll see all. Soochow gardens rank foremost mainly because of their historic background, high quality and great quantity. As far back as in the fourth century A.D., Soochow won fame as the city where was situated the garden of Ku Pi-Chiang (顾辟疆). Its exact location, though, could not be identified from the eleventh century onwards. It was the first best known private garden south of the Yangtze.

The earliest Soochow garden now still extant dates back to the tenth century. The longer the garden's history is, the less it resembles its original design. On account of repeated alterations, most Soochow gardens today began or were redone in the Manchu dynasty, generally after the later half of last century. Celebrated as the center of superior craftsmanship, Soochow boasts of fine brick work and carpentry, canals and roads facilitate communication, and agricultural production and trade contribute to a thriving economy, laying foundation for cultural activities. Then, the clement weather is ever favorable for horticulture, supported by an abundant water source. Induced by such propitious conditions, the landed and the moneyed used to flock thither to abide, constituting a major part of the leisure class. For their amusement, the nurseryman, the poet and painter pooled their talent for the laying out, construction and cultivation, and gardens flourished. Under similar conditions, gardens in other localities also sprang up, though in number not on a par with Soochow.

Yangchow (扬州) gardens rank next in number if not in quality. There are also two noteworthy gardens in Nanking (南京). Other cities like Changchow (常州), Taichow (泰州), Shanghai (上海), Nanziang (南翔), Wusih (无锡), each possesses one garden of ancient fame and is positively worth a visit. In Eastern China, one seldom finds a town without a garden or two, be it the property of a temple, an official residence, or a private individual. One can consider the entire city of Hangchow (杭州), in fact, an extensive garden dominated by the West Lake, easily the largest waterscape in any municipality. Shanghai's neighboring towns like Quinsan(昆山), Sungkiang (松江), Chingpu(青浦), Kating(嘉定), Chiangyin (江阴), Taichang(太仓), Wukiang (吴江), Changshu (常熟); farther south canal cites like Kashing (嘉兴), Nanzing (南浔), Pinghu (平湖), and Haiyan(海盐) on the southeast, all in the long past were, adorned with gardens which have since either been preserved to this day or neglected to the state of dilapidation and subsequent disappearance. Too numerous to mention chiefly these gardens, preceded by Soochow the garden city par excellence, establish the architectural style of all public and private traditional gardens in almost any part of China except the north region like Peking and Canton in the South.

More than forty years ago, two distinguished New York architects came to Shanghai to start their China tour. Ely Jacques Kahn(1884~1973)in 1935, and one year later, Clarence Stein(1882~1975) with his actress wife, Aline MacMahon. Soochow gardens they listed as a must and I had immense pleasure, on different occasions, in accompanying them around, and believe me, it was astonishing to see their spontaneous response to the esthetic peculiarities of Chinese landscape art even before I had time to point them out. Each trip was taken when wistaria was in full bloom. Each ended as a perfect day.

This publication lays due emphasis on garden planting, of which Soochow, more than other places, has a comprehensive collection. In preparing the draft, I consult frequently Christopher Tunnard's *Gardens in the Modern Landscape*, and Liu TunTseng's *Soochow Classical Gardens* (刘敦桢:《苏州古典园林》) and acknowledge my indebtedness.

Chuin Tung
Department of Architectural Research
Nanking Institute of Technology
Nanking, Jan, 1982

Contents

1. Garden as Painting
p. 222

2. Garden and Scholar
p. 223

3. Architecture and Planning
p. 225

4. Ornament and Furniture
p. 227

5 Rockery
p. 230

6 Planting
p. 234

7 East versus West
p. 237

8 Past and Present

    A. History p. 243
    B. Present p. 248

Soochow p. 249	**Nanziang** p. 262	**Changshu** p. 266
Wukiang p. 257	**Wusih** p. 262	**Chekiang** p. 266
Yangchow p. 258	**Changchow** p. 262	**Kashing** p. 266
Taichow p. 259	**Quinsan** p. 262	**Haiyan** p. 266
Jukao p. 260	**Sungkiang** p. 263	**Pinghu** p. 267
Nanking p. 260	**Chingpu** p. 263	
Shanghai p. 261	**Kating** p. 264	
	Chiangyin p. 265	
	Taichang p. 265	

1 Garden as Painting

When one is viewing a scroll of Chinese painting, one seldom inquires whether so large a man could creep into so small a hut, or whether a crooked path and the few thin planks which bridge a billowing torrent could carry the drunken recluse on his donkey to the opposite shore in safety. In Chinese painting, some rules or rather exceptions must be agreed upon before any esthetic pleasure can be enjoyed. The same convention in absurdities also applies to the Chinese classical gardens, which is in fact but Chinese painting in three dimensions.

If a visitor to a Chinese garden, after entering and before wandering too far, should pause (hesitation is wise, for he is embarking upon something not unlike an adventure), and by glimpses transcend space and volume and resolve the whole into one flat surface, he would be thrilled to realize how closely the garden resembles a painting. Before his very eyes stands a landscape, not drawn with the painter's brush, but a pictorial composition of arbor, brook and weeping willows unmistakably recalling that familiar pattern which one associates with a Chinese painting–the same crooked into a grotto. The visitor, by the way, could well be satisfied with just this much and turn away, leaving yet unseen the landscape beyond as new discoveries and new surprises for another day. For this reason, the past owner of a Chinese garden seldom lived in it and only occasionally paid it a visit. Well worth preserving was the distance that lent enchantment.

The old-school critic maintained that only a good painter could design a good garden. Incidentally this dictum was echoed two centuries ago in England by William Shenstone when he asserted that the landscape painter was the garden's best designer. An ideal combination was found in the Tang dynasty poet-painter Wang Wei (王维, A.D. 701~761) who designed his own garden Wang Ch'uan (辋川) to beguile his years of retirement. One scholar summed up Wang Wei's genius in these words: "His poetry suggests painting, and his painting, poetry." Antedating René de Girardin's "the poet's feeling and the painter's eye" by ten centuries, such a tribute Wang Wei justly deserved.

The relation between painting and garden, as between painter and gardener, was so close that the one hardly ended when the other began. If Alexander Pope declared that "all gardening is painting", he inadvertently was ranking Kent the "canvas gardener" with Wang Wei, painter, designer and owner, besides, of a garden all in one person. To depict the scenery in his garden, Wang Wei left to posterity some incomparable drawings which from then on served as the inspiration and model of all gardens befitting literary men. From these drawings one could only conclude that Wang Wei's garden was as inimitable as his painting, both being equally sublime. Nay, Thomas Whately went one step further, persisting in the dogma that landscape gardening was not equal, but "superior" to landscape painting.

2 Garden and Scholar

In China's past, every garden owner strove to imitate the garden of the scholar. The rich and the parvenu spared no pains to make their city villas and country estates look scholarly and refined. They would feel greatly flattered if they were commended not for their opulence but their taste. Such a garden as a sign of "Conspicuous waste", served well to enhance its owner's status in cultivated society and the dolce-far-niente crowd, besides providing him with a refuge in his "Rus in Urbe" (城市山林) from mundane worries and everyday struggles. Even the emperor, mighty and magnificent though he was, felt sometimes the urge to flee from his city palace, in order to live the life of a country gentleman of leisure in one of the imperial garden estates. Here he imagined himself to be Wang Wei, or some other poet, painter or recluse, and indeed, could hardly resist the temptation to adopt a "nom de plume". A different condition was to be found in the seventeenth century in France, where the "Garden Monarch" used Versailles not for seclusion and meditation but for the most elaborate entertaining and amusement. Decidedly, compared with the vastness of Versailles all other palace gardens of the world appeared cramped, and Louis XIV could hardly be blamed for his flight to the country. In the

Chinese garden, on the other hand, be it imperial or humble, crowds were not only out of place but also out of the question. Its intimate quality hampered the flow of traffic and defied the presence of multitude.

A unique feature in the Chinese garden is its association with the literary realm. No building in the garden is complete without tablets or plaques bearing inscriptions composed and written by well-known poets and scholars. Such inscriptions call for skill both in wording and calligraphy, and are constantly found in the hall, the pavilion, or over the gate-way. Every building is invariably christened an individual and apposite name or title. A parallel case existed in eighteenth century England, when poet-gardener William Shenstone put up in his own estate "Leasowes" plaques inscribed with suitable verses expressing sentiment appropriate to the scenery. As the principal leader of the romantic movement, he influenced his followers to exploit the picturesque garden to the extent that "every folly must have a name". Unavoidably the inscription excites in the visitor a literary response which combines visual pleasure with philosophical allusions. If the Chinese garden is more than painting, if indeed it is poetry or even, as was put by Lafcadio Hearn, fantasy, then these decorative writings serve the very purpose of rousing in the scholar that poetic imagination so congenial to his heart.

The scholar, and not the horticulturalist or the landscape architect, could well manage to design a classical Chinese garden any time. As an amateur, he might accomplish this poetic and romantic undertaking if not with distinction, at least with tolerable taste. Taste, be it emphasized, counts here much more than mere know-how.

The Chinese garden, to the scholar, is nothing but a wonderland of fantastic dreams that come true and a little world of make believe, and can thus be truly called an art of deception. We would not go so far as to say that the visitor is perfectly willing to be deceived. The question of reality does not bother him, as soon as he ceases to be in the garden, but begins to live in the picture. If an Oriental philosopher was not troubled by the inaccessibility to this arbor or that hill in a painting, he surely seldom found it imperative to demand otherwise in his garden. Chinese traditional painting, by the way, is far from being objective. A good painter painted not what appeared to him to be, but painted what he thought it ought to be. Stranger still, a fact entirely incomprehensible to modern mentality, was the absence of any foot path in Japanese garden before the sixteenth century. The "stroll garden" came into existence only later. The observer was simply too contented viewing the landscape as painting from a fixed point on the veranda. Hard as it might be to one's mind's eye to acquiesce to the idea of a pathless garden, imagine the horror of today's landscape architect of the Western school whose client demands a garden wherein plant and water are entirely taboo. Yet the

Japanese gardener did possess the genius of designing the "Kare sansui" (dry landscape), in which he fashioned cascades and pools without using a single drop of water. One canon in the Chinese garden design was that one had to find, with imaginative response, in the little the great, and in the evident the intangible. It was Buddhistic teaching that brought forth the Japanese dry garden. With this Zen doctrine fits in well Blake's conception–to see the world in a grain of sand and heaven in a wild flower.

A similar Oriental outlook demand the spectator to rise to the occasion in which contrast and comparison dominated the scene, and to subscribe to the idea of Robert Fortune, who visited China to collect plant specimen in 1842, pointed out that to understand the Chinese style of gardening, it was necessary to see an endeavor to make small things appear large, and large things small. After all, the universe is so vast that any garden, however extensive, is at best microscopic as an imitation of nature. In this connection, Samuel Johnson showed great sagacity to opine, in tune with the Chinese (and Japanese) philosophy of minimalism that "a little of it is very well". To exaggerate the scope of a garden, for contrast, the entrance door is made on purpose inconspicuous and casual so that a visitor can slip in without ceremony. On the other hand, the gate to the European villa is often such an artificially embellished and imposing affair that an oriental, far from feeling to wake up from a pleasant dream, would indeed wonder if he is coming back to nature after taking leave.

3 Architecture and Planning

Unusual indeed it is when one finds a Chinese garden without architecture, and by architecture we mean buildings tastefully designed and appropriately disposed. This constitutes the most important criterion of a good garden. The French paysagist Jean C.N.Forestier (1861~1930) held such view, asserting "un beau jardin avec une belle maison c'est bien rare". Ubiquitous but not to be despised is unquestionably the arbor or gazebo (ting, 亭). This toylike building can even stand on a single post, or take the plan form of a triangle, a square, a circle, any polygon, double square, interlocked circles or a cross. Its top covering thus ranges from a simple pyramid to multi-pointed and multi-hipped roofs. Upturned eaves, curved ridges and other architectural irregularities do much to lend life to the garden's already playful spirit. One or more than one sizeable structure, the pillared and rather lofty hall, occupying a key position, is the accent of a garden composition and preferably open on all four sides with removable windows to command favorable views. Such a hall usually has its raised terrace or roomy pavement adjoining the facade. A den, on the other hand, is best given a secluded spot.

Another feature in garden architecture is the covered veranda, occasionally two-storied. This colonnaded corridor mainly serves to connect building with building, or if standing between two courts, to divide yet unite them by virtue of its being open and unobstructive. If it is desirable to separate two courts entirely, the veranda is simply walled up. When the veranda happens to stand on water, it assumes the form of a covered bridge. A long veranda is not built straight but made zigzag or wavy, and if on hilly ground, must follow the contour by sloping or stepping up and down its floor. A garden building also functions as a vista or center of attraction, especially when enhanced by trees, flowers or some other ornaments. In a group of buildings, their positions are determined by comparative importance and determine, in turn, the spaces in between. Rhythm and harmony must be the chief consideration in such layout. One drawback should not be overlooked: too many buildings plus muddled arrangement lead to claustrophobia.

Garden architecture also includes, akin to the English "folly", the stone boat or "dry barge". It is a playful imitation of the house-boat, standing motionless on the lake, usually close to the shore. In the stationary cabin from the window one can look out at the scenery over the water. The stone deck provides space for chess-playing or drinking tea.

A garden might be built on the slope of a hill. In such a case, the Chinese designer displayed ingenuity in architectural planning as great as the Italian in laying out series of terracing one higher than the other. The Chinese used the terrace to provide a shelter directly below and for planting or promenading above, and one terrace became one courtyard retaining its own seclusion. Furthermore, occasionally, taking advantage of the elevation, one might even look down into a lower neighboring garden, if any, or enjoy a distant view of the surrounding countryside with some landmark like a temple or a pagoda, thus being rewarded with a "borrowed scenery" (借景). As a result, the picturesque range of the garden seemed enlarged many times. This was favorite theme adopted whenever the designer had a chance. The Japanese "shakkei", a term similar to the Chinese, achieves the same effect. One is tempted to recall the striking view of Brunelleschi's dome seen from the Boboli Gardens, or of St. Peter's from behind the fountain on

Villa Medici's terrace.

A body of water, not too large, is a desirable attribute of any garden, especially when mandarin ducks (aix galericulata) swim on its surface and carp and gold fish below. In this connection, recalling that swans signify happiness in the Moorish garden, one may also mention the white stork, symbol of longevity and companion of the recluse. The Japanese introduced from China all these creatures to enliven their garden. For lotus, it is advisable to set a limit to its growing scope so as to leave sufficient water area for reflection. An islet in the lake, with a skiff floating near by, is always an attraction. The shore might be either of earth, stone, rubble masonry, or dotted with rockery.

4 Ornament and Furniture

The Chinese garden is usually enclosed in high walls to be isolated, like the medieval pleasance, from the outside world. Different courtyards, too, when not divided by buildings, are by a wall, often with veranda along its one or both sides. No less an ornament than a functional necessity, the wall is never a stark, plain masonry structure but is usually curved in plane or undulating on top, enriched with fancy adornment (fig. 1~2). The wall surface rids itself of monotonous blank by being dotted with inlaid calligraphic stone tablets and grilles or traceries of exquisite patterns, to be constructed with thin brick or tile. The beauty of the tracery is much enhanced by its depth when sunlight plays upon it to produce the most sparking spectacle. The same tracery, by the way, may look quite different under light coming from different angles (fig. 3~4). Besides, same as when looking through the Moorish clairvoyance, one catches sight of a fragment of another courtyard or a scenic spot set in one of the framed wall openings, to thus extend space far beyond. Just like collage in modern architecture, this aperture in the wall enables one to see on both sides landscape feature overlap and interpenetrate, resulting in what may be called collage in garden art.

A pavilion or an arbor may lean upon the wall, which in turn, as done in some garden, can terminate abruptly, to be

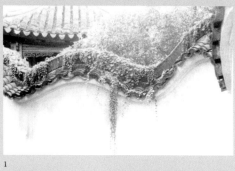

1

3

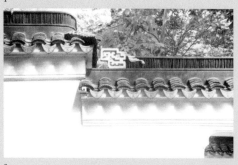

2

4

5

continued by a screen of rockwork (fig. 5). Furthermore, a wall whitewashed serves to "print" the shadow of a plant like bamboo in sunlight or as background for a piece of rockery or a quaint tree. White walls, grey roof tiles, green foliages and painted woodwork make up the dominant colors in the garden. Since the Chinese garden is introversive, it largely depends upon the wall to conceal its beauty, luring the wanderer on to a glimpse through a doorway or a tracery. The doorway is usually shaped like a full moon, a vase or a flower petal. Its design is as limitless as the wall tracery.

Pavement in the Chinese garden is also a highly decorative art and is made doubly fascinating through the use of insignificant, or even waste material. Chips of stone, broken tiles, pebbles and fragments of porcelain make mosaics inexhaustible in composition and color. Usually the design is symmetrical, in the shape of combinations of polygons or quatrefoils. Of the asymmetrical forms, the most common is the "broken-ice" pattern. Incidentally, a floor mosaic in Villa d'Este bears striking resemblance to the Chinese pavement in spirit if not in form (fig. 6).

Garden furniture outside or inside the building, constitutes the last but not an insignificant item as useful ornament. In the open air, tables and seats are primarily made of carved stone or other durable material, while interior equipment as a rule is constructed of choice hard wood. For decorative illumination, lanterns are hung from the ceiling. But curiously no care is taken to light the garden ground

6

8

7

9

Fig. 1~2 The wall is enriched with fancy adornment
Fig. 3~4 Tracery in I Yuan. Different time different looking
Fig. 5 Where wall ends and rockery begins (I Yuan, Taichang)
Fig. 6 D'Este pavement
Fig. 7~9 Garden furniture outdoors
Fig. 7 The carved stone railing (Chueh Cheng Yuan, Soochow)
Fig. 8 Bell teat stone (Hsiao Chin Shan, Yangchow)
Fig. 9 The carved stone table and seat (Liu Yuan, Soochow)

图1~2　园墙因浪漫装饰而增色
图3~4　亦园漏窗，不同时日之不同景象
图5　园墙终止而叠石开始（太仓亦园）
图6　埃斯泰别墅之铺地
图7~9　园林户外家具
图7　石雕护栏（苏州拙政园）
图8　钟乳石（扬州小金山）
图9　雕琢石桌及石凳（苏州留园）

itself at night. The Japanese stone lantern, originally a Chinese temple accessory, is here singularly absent. For diversion and augmentation of interest, tray landscape and bon-sai are placed here and there. There are also many instances of railings of carved wood or stone in exquisite workmanship for the open pavilion, gazebo, or on the water edge (fig. 7~9). But the most worthy garden ornament is, without doubt, rockery.

5 Rockery

Rockery is a queerest object only peculiar to the traditional Chinese garden. Half natural and half masoncrafted, it serves admirably well as a link and a transition between artifact and natural planting, between static architecture and verdure with life impulses. Rockery here, however, in no sense resembles the boulders which Cicero found in situ in his Roman villa and admired in writing, nor the kind of stone Thomas Whately assigned as ornament in the picturesque garden in eighteen-century England (excepting, of course, some tufa arch and grotto in Surrey). Most rockery was brought from a distance, sometimes hundreds of miles away, to its final site. One of the highly prized varieties was the hydraulically transformed "lake rock" (湖石), quarried from deep water, where on the stone through centuries of continuous washing and scouring by strong turbulence holes were bored and surface wrinkled (fig. 10~15). The Lake Tai Hu (太湖) is in the neighborhood of Soochow, consequently this garden city accumulated a large amount of supply which made rockwork facile and economical and the town itself a famed center of attraction.

It was through lake water action that rockery was reduced to look characteristically spare, porous and grotesque, strangely resembling contemporary sculpture by Moore or Noguchi, with its abstract outline and positive and negative volumes. As a unique ornament, a piece of rockery can stand on a pedestal in the manner of European statuary, or, if more than one piece, rocks may be cemented together to symbolize caves, tunnels and mountain peaks. Some rockery was so extensive that it occupied a large portion of ground as the dominant theme, like the famous historic garden "Ten Thousand Rocks" (万石园) in Yangchow. Another example is the "Lion Garden" in Soochow, still standing today, where the far-reaching but rather confused stone pile once had the misfortune of being jibed "a stupendous heap of slag".

Many a "Catalogue of Stones"(石谱) was published in China's past by connoisseurs of rockery. As a valuable document, the *Catalogue Cloudy Forest* by Tu Wan (杜绾) was reprinted by Berkeley University of California in 1961. Drawn and engraved by eminent artists and accompanied with

10

11

Fig. 10~15 Lake Rock
Fig. 10 Hsiu Yun Feng (Liu Yuan, Soochow)
Fig. 11 Kuan Yun Feng (Liu Yuan, Soochow)

图10~15　湖石
图10　岫云峰（苏州留园）
图11　冠云峰（苏州留园）

scholarly comments, such catalogues attempt to demonstrate graphically the salient features of famous rockeries in different localities through different ages. Tu's catalogue listed 116 stones. Ancient scholar's standardized criterion for a piece of priceless rockery was condensed in at least four characteristics: perforation (漏), pervasion (透), lank (瘦) and rugosity (皱). Their emotional attachment to the mysticism, even "personality", if not to the abstract beauty of rockery, was such that they seemed all but insane. This was mainly on account of stone's solid constancy which human character, alas, too often lacks. Perhaps Shakespeare sadly

Glimpses of Gardens in Eastern China | 231

12

13

erred when, in Antony's speech, he denied stone's intelligence and sensibility? Moreover, stone, being speechless, earned the scholars' esteem for its golden silence, a quality regarded as most humanly agreeable (石不能言最可人). The Bard was equally uniformed as to the Chinese scholars' devotion to, or even worship of rockery. Four hundred years before Shakespeare's time, one Chinese scholar-official had the audacity to fraternize with a piece of rock which caught his fancy but also caused his political demotion. Hundreds of years later, another literary man paid homage to a divine looking stone, and yet another was so fascinated with a miniature rockery

Fig. 12 Shui Yun Feng (Weaving Prefecture, Soochow)
Fig. 13 Shui Yun Feng (Liu Yuan, Soochow)
Fig. 14 Lake rock (Chan Yuan, Nanking)
Fig. 15 Lake rock window (Chiao Yuan, Taichow)

图12　瑞云峰（苏州织造府）
图13　瑞云峰（苏州留园）
图14　湖石（南京瞻园）
图15　湖石窗洞（泰州乔园）

14

15

that he could not go to bed to sleep without this inseparable companion.

Rockery design in the traditional garden was considered the most subtle and difficult task. Only a few so called "masons" in history could acquit themselves well. The most famous designer was Chang Nan-Yuan (张南垣) in the seventeenth century. He looked down upon the idea of imitating a natural hill with a heap of stones, but caring only for the casual, the unconventional, and succeeded in emphasizing the essentials or suggesting the existence of a hill, with a minimum amount of rockery. The succeeding century saw Ke Yu-Liang (戈裕良) confining his labor to making a rock hill, usually moderate in size, well, with elegant relish and spiritual transcendency. One can still see today a sample of his matchless achievement in Soochow, in the small garden Huan Hsiu Shan Chuang. Here he improved the ceiling in grotto construction with vaulted stone work instead of the time-honored usage of flat slab. He was considered the first and foremost expert in the art of rock-hill making by scholars contemporary and after his time.

But the most remarkable and enormous in extent was the ancient imperial rock park built by Emperor Hui Tsung (徽宗, A.D. 1100~1126) of Sung dynasty, al-

Glimpses of Gardens in Eastern China | 233

though history dates back the first appearance of rock hill to the first century B.C. The Emperor being an inimitable painter by his own right, lavished his attention much more on rockery than on statecraft. Barges swarmed on the Grand Canal, loaded with stones from Tai Hu, to build his favorite rock park "Ken Yao" (艮岳, "Northeast Summit") in the palace ground. Came one day the Tartars from the North to besiege the capital. Much rock work had to be broken into canon balls to defend the city before the emperor became captive himself. Choice rockery pieces were soon carted to Chung Tu (中都, "Inner Metropolis") by the Tartars to adorn the hill on an island, in the Imperial Park now known as Peh Hai(北海), in Peking.

6 Planting

Past writers on Chinese gardens seldom gave adequate account of plant life, not that they regarded verdure with indifference but because flowers and trees, from their standpoint, played only a subordinate and complementary role. It was no surprise that a similar standpoint reduced flowering trees in the Japanese garden to the minimum. The function of trees and flowers is to match the environment in an organic way, taking into consideration topography, sunlight, humidity, growth habit, and characteristics of the vegetable kingdom of a given locality. Posture, pattern, configuration, color and scent establish the basic criteria for selection of a species.

Over and above such practical deliberations, plant life, besides serving to screen off those elements best not seen too soon, also affords material for topics of poetry and painting, common specimens being pine, bamboo, and plum, the three proverbial congenial companions in wintry weather (岁寒三友). An old tree is ever prized for its venerable age and always invaluable as a welcome bequest to a new garden rebuilt on the same site.

One must remember that although the Chinese garden is an intimate, human, and sophisticated art, yet plant life is usually free from any appearance of artifice.

Here you find neither trimmed hedges, nor flower beds arranged in geometrical patterns. Topiary, naturally, is ruled out. Whatever fantasies the European landscape architect lavishes upon plants, the Chinese gardener reserves to embellish his bowers and follies. Most conspicuous is the absence of the mown and bordered lawn, which, though attractive to the cow, has but little to appeal to the human intellect. Geomancy, and even superstition often left existing topography with attached objects undisturbed. Much of nature is scrupulously let alone, yet the Chinese garden never looks botanical. Dr. Johnson, who confessed he could not tell a cabbage from a cabbage rose, dogmatically queried Boswell, "Is not every garden a botanical garden?" If the great lexicographer's definition was final in the English sense, in Chinese usage "botany", dealing with healing herbs, had only to do with the apothecary and the invalid. Evidently his good friend, Sir William Chambers, who saw and admired the Chinese garden neglected to convey to him one of the essential qualities of this branch of Oriental art. Trees and flowers were never made to outline other garden features but always had their place and purpose and sometimes, even preeminence. The *Paeonia suffruticosa* in Loyang (洛阳) in the twelfth century and the *Paeonia lactiflora* in Yangchow in the eighteenth enjoyed such fame that they immortalized the gardens they adorned. But even in these cases flowers and trees, well chosen, were made to appear casual and unobtrusive. The English landscape garden, however, took the hint and went to the extremes by abolishing flowers altogether in the "capable" hands of Lancelot Brown, when he was given a chance to "improve".

If the Chinese garden is never meant to be monumental, plant life plays not an insignificant part in its contribution to fulfill this purpose. Here one finds no majestic cypresses to make an avenue, or rond-point and gushing fountain to jet forth to a certain height. All these seem to be, to paraphrase Wilde, "two touches of nature". Also unthinkable is the major and minor axes in garden layout like the West, where the arrangement which can only breed an awe-inspiring and desolate spectacle. As a consequence, the Western garden never rids itself of the look of a wilderness of vegetation. In the tradition of Le Notre verdure always dominates the European garden so much, so that if allowed to grow unchecked it would run wild and eventually revert to a state of "forest primeval". Well known was Le Notre's emphasis on unhampered openness of Nature and he did not tolerate anything that restricted the vista. But it is time now to examine the relation between planting and other garden artifacts.

When the garden court is spacious, a big, solitary tree, preferably of antique variety, can serve as the main theme of a composition as well as the unifying factor between different buildings. Besides, as a pivot, it defines major and minor court spaces and ever once so often lends its poetic name written on plaque to a building or a compound. On the other hand, on narrow ground, where sightline is restricted, a sizable tree has no place, but rather group planting, with pleasing configuration, rich in hue and perfume, supplemented with perhaps a lone piece of rockery, is invariably the rule.

In mixed planting deciduous trees ought to dominate, supported by evergreen and strengthened by clumps of bamboo, encompassed in turn by creepers and grass plants. Forceful trunks and strong branches best stand side by with pliant ones, and deciduous foliage in big mass may combine with evergreen showing sparse vs. dense growth, light vs. dark color, changing with the seasons, avoiding monotony and repetition. Thus in spring time appear magnolia and peony; there are *Lagerstroemia indica* and lotus in summer; autumn is the season for chrysanthemum, maple; and *Chimonanthus praecox* and camellia are in full bloom in winter. Most ideal is to see each season consummate its own outstanding blossoms by turns in one garden court all the year round.

Favored by nature, Soochow grows almost any plant peculiar to the north temperate zone. In this city in any big garden one can count more than a hundred species, while in medium and small garden, from twenty to seventy kinds. Lotus, wistaria, plum, osmanthus, begonia, jasmine and chrysanthemum are common to all gardens. No garden is complete, of course, without bamboo. The usual evergreens are pine, juniper, and cedar. There is an almost endless variety of trees-willow, maple, *Firmiana simplex*, palm, *Musa basjoo*, elm, and so on. Well known is the ingenuity of the Chinese nurseryman, since time immemorial in transplanting and cross-breeding flowers and trees. Plants in Chinese gardens deserve an independent treatise by scientists in physiology and can be roughly divided into the following types according to their sensual or utility values:

1. Flower type. The most splendid spectacle in the garden is the blossom plant excelling in color or scent or both. In the evergreen category there are camellia, magnolia, rose, *Osmanthus fragrans, Rhododendron, Nerium indicum, Gardenia jasminoides, Serissa serissoides, Michelia figo.*, etc. Deciduous are *Paeonia suffruticosa*, plum, peach, almond, malus, lilac, hibiscus, and so on. Peony is the indisputable crowning glory on the flower terrace of any garden by virtue of its brilliant color and extraordinary size. Camellia and Prunus mume are especially welcome in the cold season when other plants cease to blossom.

2. Fruit type. From early summer to autumn, some trees begin to bear fruit, good to look at and mostly edible, such as the golden loquat, citron, and *Nandina domestica,* fig and date trees are deciduous.

3. Foliage type. This type is indispensable in any garden. For all the year round ornament there are the boxwood, yucca and palm. Though weeping willow, tamarisk and maple shed their leaves in winter, they add variety in color and carriage

in other seasons to the greenery of the woods, especially the maple, whose scarlet splendour constitutes an outstanding adornment in autumn.

4. There are also trees which, essentially as component part of fundamental planting, above all afford desirable shade in hot summer, such as the juniper and conifers, together with the deciduous ginkgo, elm, locust and especially *Pterocarya stenoptera*, which, leafy as well as branchy, easily can create surroundings of luxuriant atmosphere.

Not to be forgotten are the climbers and creepers such as the evergreen rose and ivy, and the deciduous grapevine, wisteria and morning glory, all serving to either climb up trellises as decoration or to fill up voids, cover bare surfaces of building element or certain rockwork. Then there are plants which contribute to the pleasure of listening, like the rustling of leaves in wind, rain drops on a family of lotus, and the chirping of birds and insects. Besides, there are numerous species of bamboo and grassy growth, enriched by aquatic plants like lotus and lily, which are in turn surrounded by reed and rush near the water shore line to lend that marshy touch to the immediate environment.

7 East versus West

A French poet once declared: "J'aime fort les jardins qui sentent le sauvage". This just hits upon the difference between Western and Chinese gardens, the latter being entirely devoid of the jungle atmosphere. The Chinese garden is primarily not a single wide open space, but is divided by corridors and walls into courts in which buildings, not plant life, dominate the scenery and attract one's attention. Garden architecture in China is so delightfully informal and playful that even without flowers and trees it would still make a garden. This is especially true in Japan, where in the Ryoan-ji Garden, Kyoto, there is absolutely no plant life, only stone, sand

and a wall of rammed earth being present. For compensation, however, its saving grace lies in the thick grove immediately outside the boundary. On the other hand, Western gardens consist much more of landscape than architecture, to such an extent, buildings, if any, stand in splendid isolation. Foliage, flower and fountains are more akin to one another than to the casino and kiosk, in spite of the effort to arrange verdure and jet architecturally, even to the extent of laying them out symmetrically in axes.

William Chambers was the first Englishman to seriously study the Chinese garden, and in his *Dissertation on Oriental Gardening*, tried to prove the superiority of the Chinese landscape art (fig. 16). He had the good fortune of coming to China just during the reign of the Manchu Emperor Chieh Lung (乾隆, 1736~1795), the golden age of Chinese garden art, while it is absorbing to debate on the relative merits of Chinese and European gardens, it is futile to claim which is better, so long as each is harmonious with the art, philosophy and life of its respective worlds, each is as great as the other.

Standing halfway between East and West is the Alhambra of Granada. This Mohammedan castle gathers halls and verandas around each of its many courts, kindred in spirit with the Chinese practice, but with verdure and water laid out in geometrical fashion. Symmetry is evident, yet the haphazard and intimate arrangement savors none of the rigidity and monotony of the Western garden.

Italian villas, inspiring the classical gardens of the West, excel especially in terracing and exuberant vegetation, provided generously by topography and climate. Parterres, marbles, staircases and cascades are arranged in strong formality. Tall cypresses stand in formidable rows. But the Roman garden, with all its axes and repetitions, by virtue of its successive levels, does achieve one object–to surprise, which is incidentally one of the reasons for which the Chinese garden is designed. Ascending the steps in the terraced Italian villa, one stop on a platform and looking below, suddenly comes upon that superb view which is incomparable. In almost all the Chinese garden courts, the visitor has a similar experience. But here instead of a whole panorama at a glance, he sees a mere fraction of the ensemble, only to be surprised again and again as he explores further on. Worlds open out to him, verses and inscriptions carry away his imagination, and vistas tempt his curiosity. This element of surprise is singularly absent from gardens in France, a nation of rationalism and logic, where flat terrain and vast expanse of the parc heightens the scene of monotony.

But the Chinese garden has its own irrationalities which must seem ridiculous to the Western mind. Who should suspect that the upper floor of a pavilion is often inaccessible? One is lucky to find a workable ladder! Then there is the narrow and serpentine footway which covers the longest distance between two points, and the slippery, almost perpendicular rock hill so

Fig. 16 A private garden in the eighteen-century (Canton)

图16　18世纪的私家园林（广东）

precarious as to discourage climbing. A meandering stream winds its way under the low and zigzag bridge, whose function, oddly enough, seems to conduct you near enough to the water to be drowned rather than to cross it. Confusion worse confounded! Those who are impressed by the monumental beauty of the Italian villa and the simple charm of the English park cannot but feel perplexed at such incautiousness and absurdity.

All these unpredictable features belong to a school of thought entirely compatible with the philosophy of the Chinese scholar, and did not Kent concur with such philosophy when he asserted that "nature abhors a straight line"? If direct walks, long avenues and well balanced parterres are results of what Kenneth Clark termed the "measuring frame of mind" of the mathematical attitude of the West, the Chinese garden, an immeasurable art, avoids at all costs any stiff orderliness and geometrical rigidity. In it curves and studied irregularity known as sharawadgi or beau disorder characterize the design, and space disposition limits visibility to one single pictorial courtyard only. To play to the full on the hide-and-seek motif, the visitor's movement in the puzzle would be ever so often deviated and sidetracked. But it matters little. Is it not so much more enjoyable to travel than to arrive?

To the decadent in the extreme, a pleasure delayed is a pleasure twice enjoyed. The garden's kaleidoscopic composition conduces to human relaxation and random progress. Attention is constantly attracted by diversity of interest. Intricacy and complication are further enhanced by contrast through open vs. closed space, dark vs. bright spots, high vs. low openings, and big vs. small surface or volume. To create myriad vistas and various centers of interest, not only walks are curved, but the ground is often irregular in contour to interrupt sight line so that vision is confined to a little at a time. Not so the Versailles. Its open arrangement and far reaching view became so tiresome that maze and labyrinth had to be invented, and to compensate for linear monotony, little secret bosquets were plotted amongst groves to beckon the curious and the wayward. One cannot but conclude hence, that if the Western garden only pleases the eye, the

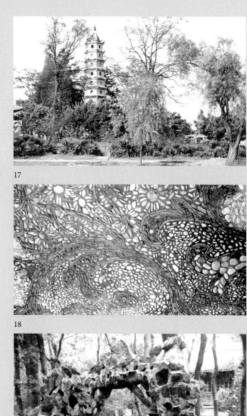

Fig. 17~22 Certain features are common between the traditional Chinese garden and the eighteen-century English pictorial landscape. Never the twain shall meet? Kliping might sing otherwise. The similarity is striking
Fig. 17 A park in Nanking
Fig. 18 Pavement in Ho Yuan (Yangchow)
Fig. 19 Rockery arch in a temple garden, Nanking
Fig. 20 Kew Garden, England
Fig. 21 Pavement in Generalite, Spain

Fig. 22 A tufa arch in an English garden

图 17~22　某些特征为中国古典园林与18世纪英国如画景观所共有。难道两者不应相遇？开普林或许要换一种吟咏方式，相似之处 是如此惊人
图 17　南京某公园
图 18　何园铺地（扬州）
图 19　某寺庙之石券门（南京）
图 20　英国邱园
图 21　简奈莱利特铺地（西班牙）
图 22　英国某园林之凝灰岩券门

Chinese garden aims at pleasing the mind. The one relies on measuring rod, the other communicates with intellect. In the final analysis, it is a case of mind vs. matter.

After the Chinese garden was finally completed, in the early years, while architecture and other artifacts were soon mellowed with age, some planting still lagged behind; and when at last trees had grown old gracefully, buildings were approaching disrepair. Rockery, nearly indestructible, endured longest. This problem was particularly annoying as classical garden in China, not like those in the West, leaned preponderantly on architecture. But an Oriental philosopher viewed such vicissitudes with complete equanimity, his detachment was easily comprehensible since he looked at his gardens only at great intervals, just as he looked at his collection of rare old paintings. Both demanded an occasion. Both improved with age.

One gathers from this essay that certain features are common between the traditional Chinese garden and the eighteen-century English pictorial landscape. If imitation is the sincerest form of flattery, then the English romantic school paid China the highest compliment by following, either unconsciously or pari passu on purpose, the Chinese example (fig. 17~22).

Many foreign lovers of the Chinese garden, from William Chambers down, have written books on it. One or two contemporary cases will suffice. Osvald Siren's *Gardens of China*, published in 1949, dealt chiefly with what he saw in the North. A long-time resident in Soochow,

F. R. Nance, compiled in 1936 a treatise *Soochow the Garden City* (pub. Kelly & Walsh, Shanghai). While its author modestly made no pretense of writing for the scholar but rather for the sightseer, this delightful little book deals with six major Soochow gardens, all worth a visit and study. Talbot Hamlin, in his *Forms and Functions of Twentieth Century Architecture* (published in 1952) devotes a section to "Gardens and buildings", in which one finds two Soochow garden plans and an account of the curving path, variety of views, the changing landscape, the mystery and climax, all being intrinsic qualities contributing to the fantasy and picturesqueness of Chinese gardens. The Chinese garden, like any garden anywhere, is par excellence an art of peace.

Brigandage, warfare and the elements have been its chief destructive force. While an impecunious or irresponsible owner might easily neglect his garden to the point of decay even during peaceful days, in times of armed conflict few gardens could be spared from damage or annihilation. Repair and restoration work done by the People's Republic after 1949 revive this glorious art to a great extent, notably in Soochow and elsewhere. But one can never emphasize too much Soochow, where famous and less famous old gardens, big as well as small, amount to the number of over a hundred, giving the city a unique position unchallenged by any other locality.

Besides, one must not forget that in addition to violence, there is also the subtle and peaceful agent which tends to undermine the very continuance of the already hazardous existence of the Chinese classical gardens, and that is the landscape architecture of the West which is fast becoming fashionable as a school subject in modern China. The Chinese classical garden, like any antique Chinese painting and other traditional art works, is in danger of being relegate to be mere archaeological relic if allowed to go its own way uncared for. Many celebrated gardens would have gone to the realm of oblivion had not timely measures been taken to come to their rescue. A great effort has already been made by the People's Authorities to revive public interest in appreciation and reappraisal of traditional garden craft. We who have seen and studied this frail and fast fading flower of an old culture owe to posterity and the world at large a faithful and adequate account of those gardens that justly deserve preservation and enjoyment.

8 Past and Present

A. History

The crude beginning from which has been developed the Chinese garden such as we see today may be dated at about 1800 B.C. when in the Hsia(夏) dynasty king Chieh (桀) built for pleasure his "Jade Terrace" (玉台). More than six hundred years later, the first ruler of Chou (周), Wen Wang (文王), had, in addition to his "Holy Terrace"(灵台), the "Holy Pond"(灵沼) and the "Holy Menagerie"(灵囿), as recorded in the *Book of Songs* (诗经). This Book also alludes to orchards, vegetable gardens and bamboo groves. The germ of the idea of a garden undoubtedly found fertile soil when the nomad had adopted the more stable occupation of agriculture. During the reign of Shih Huang-Ti (始皇帝, 246~210 B.C.) of Ch'in dynasty, imperial gardens for the first time were built on a grand scale, combining "Terraces" and "Ponds" with woods and menagerie into one pleasure hunting ground. The park known as "Shang Lin" (上林, "Imperial Grove") was much enlarged by Emperor Wu Ti (武帝, 141~87 B.C.) of the Han dynasty, and with other parks and villas embellished the metropolis Ch'angan (长安), now Sian (西安), capital of Shensi province. The limited span of an emperor's life, unfortunately just as inflexible as that of any common man, greatly distressed the monarch. All he could do was to follow the advice of a wizard-priest and to erect artificial hill surrounded by water, symbolizing the Elysian paradise where he hoped to attain immortality. From this theme sprang the island-in-pool idea in Chinese and later, by transplantation, in Japanese garden design.

The garden by then, however, was no longer the sole play thing of the Son of Heaven. The princely villa of Liang Hsiao-Wang (梁孝王), Emperor Wu Ti's uncle, in the province now known as Honan (河南), boasted of a rockery hill, which records prove, was his invention. In about the same period a multi-millionaire Yuan Kuang-Han (袁广汉) indulged in similar pleasures, his garden being near Loyang, also in Honan. The reign of Wu Ti, too, witnessed the earliest scholar's garden of Tung Chung-Shu (董仲舒), who, it was said, in order to concentrate his mind, pulled down the curtains for three years, so as not to be distracted by the garden scenery. The scholar's garden in those days, being modest, could not have meant more than planting a few trees and flowers in the courtyard, similar, no doubt, to the medieval cloister garth in Europe.

The Tsin dynasty (A.D. 265~420) first saw the popularity of private gardens. Shih Chung (石崇), a shipping magnate as well as a scholar, built his Val D'or (金谷园) in Hoyang (河阳), again in Honan. Here his extravagances know no bounds, One of his favorite pastimes was to make fair damsels walk over aloes powder. If any one left so much as a slight foot print, she was considered too plump and was at once ordered to diet. Here his beloved mistress "Green Pearl" (绿珠), like Messalina, paid for her passions with her life. According to Shih Chung's own account, his "Val" consisted of trees, lakes, pavilions and towers. Birds and fish abounded. He was entirely occupied with angling, music and reading. More than half a century later, Ku Pi-Chiang (顾辟疆), as scholarly as he was aristocratic, had a garden in Soochow, first of its kind ever to appear in Eastern China, famed and admired during succeeding generations till Tang dynasty. In his own days his garden was the envy of natives and tourists alike. Once he had a violent quarrel with an intruder as arrogant as himself, the unwelcome stranger being no less a person than the heir of Wang Hsi-Chih (王羲之), both great calligraphists father and son.

During the golden age of Tang, gardens and villas crowded the capital Ch'ang-

23

24

Fig. 23~26 Wang Ch'uan (Wang Wei)
Wang Wei's roll-painting of the twenty scenes of Wang Ch'uan. There is a replica of it made by Guo Chung-Shu of the Sung dynasty. The present replica is made by Guo Shi-Yuan in 1617, the year Wan Li 45 of Ming dynasty. It has five parts which includes scenes between Huazi Mount and Bamboo Alley House
Fig. 23 Wang Ch'uan: The Huaizi Mount
Fig. 24 Wang Ch'uan: The villa at

25

26

the mouth of Wang Ch'uan
Fig. 25 Wang Ch'uan: The north pile
Fig. 26 Wang Ch'uan: The south pile, Bamboo Alley House

图23~26　王维辋川二十景画卷，有宋郭忠恕摹本。此摹本为明万历四十五年，即公元1617年，郭士元所绘。画为五段，从华子岗到竹里馆
图23　辋川：华子岗
图24　辋川：辋口庄
图25　辋川：北垞
图26　辋川：南垞，竹里馆

an and its suburbs. Thither politicians repaired to escape the summer heat. Gardens of scholars and artists were numerous, the most celebrated one being the Wang Ch'uan of Wang Wei, poet-painter. This estate vast in extent depended much on natural scenery for its charm. Its renown was augmented as mentioned before, by his painting of the entire layout, consisting of twenty sceneries, which were faithfully copied by later artists (fig. 23~26). Another Tang poet, Po Chu-Yi (白居易), acquired the habit of building gardens wherever he stayed, even when his sojourn was of short duration. The fruit of his labor could not be very elaborate, but it was enough for him to see a hillock and

Glimpses of Gardens in Eastern China | 245

a streamlet in order to feel the nearness of Nature.

Many private gardens were built in Loyang in the Northern Sung dynasty (A.D. 960~1127). Li Ke-Fei (李格非) gave an admirable account of them in his *Celebrated Gardens of Loyang* (洛阳名园记), in which over twenty-five gardens were described. He stressed the importance of restraint in garden design, pointing out the disadvantage of over-sized water pond and too much artificiality. It was his sadness to see how "fortune's fickle wheel" accidentally determined the existence of gardens, here today and gone tomorrow. Kaifeng (开封), the capital city, was also famed for its gardens, lorded over by the imperial rock park, the Ken Yu.

There were two garden cities in the Southern Sung dynasty (A.D. 1127~1279), Hangchow (杭州) and Huchow (湖州). Hangchow then the capital, besides numerous private villas was adorned with many imperial parks. The incomparable beauty of this city lies in the famous West Lake "Hsi Hu", made doubly attractive by surrounding hills. Huchow, known as Wu Hsing (吴兴) today, possessed no less than thirty gardens, none of which are now traceable. At present, in these two cities, we find only mediocre gardens wholly unworthy of their past glory. In another garden city, Soochow in Northern Sung dynasty, two gardens were especially well known. Chu Mien (朱勔), the rockery-commissioner for Ken Yu, to hunt for the required material, literally left no stone unturned to fulfill his duty. For his own villa in this city, he exhausted his talent on it, naming it "Green Water" (绿水园), long since disappeared. Another garden was the "Pleasure Orchard" (乐圃) built by Chu Peh-Yuan (朱伯原), and though altered many times so that it is almost beyond recognition, still stands today as "Huan Hsiu Shan Chuang".

Private gardens flourished in full glory in the Ming dynasty (1368~1644), chiefly in Kiangsu (江苏) and Chekiang (浙江) province. Notable centers scattered mainly around Soochow. Quite a few of Ming gardens, after many vicissitudes, have survived to this day. Garden making became a special, if not somewhat mystic, branch of knowledge and was timely dealt with in a treatise named *Yuan Yeh* (园冶, "On Garden Making"), published in about 1634, by Chi Ch'eng (计成). In this unique book, he set forth theories and practices and described various branches of landscape art, supplemented by chapters on decoration, with illustrations. He did not dwell on flowers and trees in a separate chapter but simply mentioned them here and there, putting them in the place they deserved. A contemporary of his, Chu Shun-Shui (朱舜水), seeking safety in Japan during the Manchu conquest, did much to leave a strong Chinese influence on Japanese garden design, especially the garden of the scholar. His accomplishment can still be witnessed in the Koraku-en (后乐园) in Tokyo.

In the early Ching or Manchu dynasty (1616~1911), Yangchow was transformed into the most splendid garden city history ever saw. The Emperor K'ang Hsi (康熙) and his grandson, Emperor Chieh Lung visited the city during many of their southern tours. Artist and dilettante, Chieh Lung tarried in all the pleasure gardens, wrote poems about them, and carried away the choicest rockery. The unique feature that made Yangchow matchless was the cluster of gardens along the Long and Narrow Waterway (瘦西湖), one after another, forming an unbroken chain from the city to the hill. Then was the age when money flowed freely. Boats loaded with salt from Yangchow depot sailed to different parts of the Empire, and returned with, besides cash, also blocks of rockery as ballast, which came in handy for making artificial hills at low cost. Just at this time, too, a host of nursery men came into existence, the most skillful and best known arboricultural business being carried out by the Wang family, whose descendents are even now still making a living on this craft. With easy money and easy material, any well-to-do family in Yangchow could manage to have hill-and-water design carried out in his home compound, from a mini courtyard to a sizable villa.

Most of the famed gardens were built for the sole purpose of pleasing the Emperor and coveting his August presence. No effort was spared to satisfy his whim. On one occasion when the Emperor was sailing on the canal he expressed a desire to see a pagoda near a certain temple to consummate a vista. Lo and Behold! The following day Nature was improved by the very presence of a pagoda. The Emperor was, needless to say, greatly amused. But the job was so hastily done that it had to be rebuilt later.

The Emperor Chieh Lung also included Wusih, Soochow and Hangchow in his several itineraries south of the Yangtze, and graced by his presence all the gardens worth his attention. Some of these he caused to be duplicated in Peking (fig. 27~28). Few rulers in any land, except perhaps Hadrian of Rome, travelled so extensively in his domain and built so many copies of his favorite haunts to preserve pleasant memories. After Chieh Lung's death, however, almost all garden in Yangchow were neglected, decayed, and finally disappeared. Only two or three of them have survived, notably Hsiao Chin Shan (小金山), a temple garden, and Ping Shan Tang (平山堂), in addition to a host of private gardens.

In Nanking, two princely gardens have survived to this day since the fourteenth century, namely Hsu Yuan and Chan Yuan. Both owe their preservation to being continually occupied as adjust of government offices put into timely repair.

But the best known private garden in Nanking in the early Manchu dynasty was, beyond dispute, Sui Yuan (随园) owned by Yuan Mei (袁枚), epicure, scholar and poet. He purchased a ruined villa on the slope of a hill in 1748, and devoted the rest of his life to cultivating this famous garden. Like Pliny, he was one of the very few literary men who had the good fortune of owning property for pleasure. After his death, the garden, undergoing negligence and repair

27

Fig. 27~28 The duplicated garden of Soochow in Peking
Fig. 27 Summer Palace (Peking)
Fig. 28 Chi Ch'ang Yuan (Wusih)

图27~28　北京的苏州园林复制品
图27　颐和园（北京）
图28　寄畅园（无锡）

by turns, stood until about 1853, when it was demolished during the Tai-Ping occupation.

Many gardens were destroyed during the internecine war between the Manchu government and the Tai-Pings (1851~1864). It still remains true that both in quantity and quality, before as well as after this period, gardens in the provinces Kiangsu and Chekiang stand supreme. Selected extant gardens described in the following section are all in these two provinces.

B. Present

Speaking of Sung, Ming or even early Manchu dynasty gardens, we usually have no evidence of them other than their sites. Insurrection, fire, and the perishable material with which garden buildings were

28

constructed, accounted for the rapid disappearance of so many famous gardens in the past. The Tai-Ping war gave the last blow. Whatever gardens we see today are but remnants of a bygone glory. With the introduction of glass and concrete, some of the charming features in garden building like lattice paper window and pebble pavement gradually yield ground to pane glazing and cement footway. Tragedy of progress, in which commercialism beats art, and building lot subject to urban development, is painfully evident in the fortune of the traditional Chinese gardens, one of the most exquisite achievements of mankind.

• Soochow

1. Chueh Cheng Yuan (拙政园, or Manchu Garden), situated in the northeast part of the city, was built on the site of a temple in Yuan dynasty by the Wang family in the first quarter of the sixteenth century. Shortly after, it came into the hands of Hsu, as payment to discharge a gambling debt. At the beginning of the Manchu dynasty, it belonged to the Chen family. When the Manchu garrison was expanded, it became the headquarters of the commander. In 1679, it was occupied by the Civil Administration Office, but soon passed into private possession again. In about 1742, a Chiang family completely restored it, naming it "Fu Yuan" (复园, "Revived Garden"). In the course of a hundred years afterwards, it again changed ownership twice (fig. 29). During the Tai-Ping war it was occupied by one of the insurgent leaders. After the governor's residence, but soon in 1872 became the Manchu Guild House (八旗奉直会馆), hence the garden's name. The Manchus, however,

29

30

left the property after the 1911 Revolution, and the garden was open to the public.

On the west is another garden, which, being originally part of the Manchu Garden but separated by a wall belonged to the Yeh family and then in 1877 came into the possession of Chang, until 1949. (Chueh Cheng or "Manchu Garden" means "Garden of the stupid official, who, unwise in public affairs, hopes to be an apt gardener".)

2. Shih Tzu Lin (狮子林, or Lion Garden), in the neighborhood of the Manchu Garden, was originally the property of a Sung dynasty official and later sold to a monastery in Yuan dynasty. Its abbot Wei Tse (惟则) initiated the stone hillock in 1342, which subsequently was collaborated into the far reaching rockery pile we witness today. The rock hill owes its fame partly to Ni Tsan (倪瓒, 1301~1374) who made a painting of it some forty years after its completion (fig. 30). The garden was separated from the temple at about the end of Ming dynasty. The Manchu Emperor Chieh Lung visited the garden when it belonged to the Huang family, and was so fascinated by it as to cause a garden to be similarly laid out at Jehol, and then again another, at Peking. There seems to be two explanations as to why it was called the Lion Garden. One is because of, on the stone hill, a piece of rockery resembling that very animal. Some people suggest, however, that the abbot, who came from a temple near the "Lion Cliff" at Tien Mu Shan (天目山), in Chekiang province, named his rockwork in Soochow "Lion", to remind himself of the old days. The garden was transferred in 1917 to Mr. Pei (贝), who made his millions through transaction in dye stuff. Except the rockery that did not change much, the rest of the garden, mostly finished in the '20s, could not help but put special emphasis on showy detail and meticulous workmanship that reflected the owner's opulence.

3. Liu Yuan (刘园, now 留园), whose

Fig. 29 Chueh Cheng Yuan, Soochow (Painted by Tai Hsi in 1836)
Fig. 30 Shih Tzu Lin, Soochow (Painted by Ni Tsan)

图29　苏州拙政园（戴熙绘于1836年）
图30　苏州狮子林（倪瓒绘）

origin dates back to the Hsu family, is located outside the city. In about 1794, it belonged to the Liu family, hence the name. When it was acquired by the Sheng family in 1876, the new owner, wishing to enjoy the popularity of the garden's old title yet to keep its sound unchanged, renamed it "Tarry a While (留) Garden". Being a rich mandarin, he lavished a fortune on the garden, with additions and alterations. The central portion is magnificent. On the east, numerous courts and corridors form an extraordinary maze, containing some choice pieces of rockery. The architecture in general is excellent (fig. 31).

4. Huan Hsiu Shan Chuang (环秀山庄) on Ching Teh Road (景德路) owes its origin to a princely garden in the "Five Dynasties" (tenth century). In the subsequent Sung dynasty, it belonged to Chu Peh-Yu-an, who named it the "Pleasure Orchard". A Chang family came to possess it in the Yuan dynasty. In about 1470, it was sold to Tu Tung-Yuan (杜东原), and later to a polit-

Fig. 31 Liu Yuan, Soochow (Photographed in 1931)
"Un beau jardin avec une belle maison c'est bien rare." _Jean C. N. Forestier

图31　苏州留园（摄于1931年）
"一座优美园林附带一所优美建筑，实属罕见。"——让·C. N. 法雷斯特

ical figure named Shen Shih-Hsin (申时行). During the reign of the Manchu Emperor Chieh Lung, it passed into the hands of the Chiang family. The rock grotto, diminutive yet exquisite, is much superior to the enormous pile in the Lion Garden, and serves as an example of "less is more" in Miesian parlance. The garden suffered some damage during the Tai-Ping war. Repairs were done in 1898. In the '30s, it was owned by the Wang family, who donated it, meantime, to the clan-welfare establishment, which has since been known as "Wang Yi Chuang" (汪义庄).

5. Yi Yuan (怡园, "Garden of Ease") is situated in the center of the town, on the west of the site of a Ming dynasty residence of Wu family. The main part of the present garden was done in the '90s of the last century by Koo Wen-Pin (顾文彬), a rich retired mandarin, whose descendants,

mostly scholars and painters, did the enlargement and naturally left their marks on the layout. But the result was hardly ideal. The truth is that when too many fine examples were available to copy, as in Soochow, it became a handicap and not an advantage to decide on the best choice.

The garden is noted for its collection of Sung and Ming dynasty furniture and numerous fine flowers and trees.

6. T'sang Lang Ting (沧浪亭, "Surge Pavilion") boasts of being the most ancient garden now extant in Soochow city. It began as a villa of a governor-general in the tenth century. About a hundred years later, a Sung dynasty scholar Shu Shun-Chin (苏舜钦) purchased the place and raised, on the water edge, an arbor, the famous pavilion "T'sang Lang Ting" which has since become the name of the garden. During the fourteenth century, the property was converted from a garden into a temple; and in the middle of the sixteenth century, a monk there named Wen Ying (文瑛) duplicated an arbor, which bore its previous name, thereby converting the temple back again into a garden. In 1695, the premise was considerably enlarged, with the arbor T'sang Lang Ting being moved to the hill top, its present site. Repair work was done in 1827, adding some new buildings here and there. The garden suffered damage during the Tai-Ping occupation and extensive reconstruction was completed in 1873. This garden, with its hill and sylvan environment, smacks of genuine atmosphere of the countryside, thus distinguishing itself from many other gardens in Soochow and elsewhere. In addition, to the east of the entrance gateway, stands a hall called "Facing Water" (面水轩) dispensing entirely with any boundary wall and thus eliminating the sense of division between the garden proper and the adjoining pool, and merging both together to enlarge the landscape–an ingenious device seldom seen anywhere. The numerous wall traceries, formerly totaled 108, none two alike, are also famous for their variety and charm.

7. Wang Shih Yuan (网师园, "Fisherman's Garden") began in A.D. 1174 as the villa of a retired official Shih Cheng-Chih (史正志), who called it the "Fisherman's Cubbyhole" (渔隐). Dilapidated and obscure for centuries, the premise came into the possession of Sung Tsung-Yuan (宋宗元) in about 1765, who built it anew, and named it the "Fisherman's Garden" to revive the ancient title. The place later became neglected once more, and it was in the late eighteenth century that Chu Yuan-T'sun (瞿远村) bought, revived and enlarged the garden, known then as "Chu Yuan" (瞿园). He did much to establish the layout we see today. It was afterwards sold to the Wu family and again to Li Hung-I (李鸿裔), a retired official, soon after the suppression of the Tai-Pings in about 1870. Li enjoyed the garden for quite sometime before his heir sold it to a Manchu scholarly general Ta Kuei (达桂), who owned it until after the country had become a Republic. From the Ta family another scholarly general Chang Chin-Pe (张金坡), who once had resided in Mukden as governor,

bought the garden but never came to see it. His relative, Ho Ya-Nung (何亚农) purchased it in the '30s and owned it until the '50s, when it was donated to the public.

More than one painter had lived in this garden from time to time. Its aesthetic and romantic atmosphere no doubt did much to inspire the artists in their creative endeavor.

A pond stands in the garden's center, dividing the landscape with buildings into two parts–north and south. The southern group was used by its owner for light entertainment while in the north the place was meant for study and more formal occasions. East of the garden are successive courts with halls and rooms enough to accommodate a big family. The garden's architecture is superbly tasteful and the planting skillfully chosen and well arranged. Salient features of this garden's courtyard and building in the northwest corner were reconstructed as a gift from Brooke Astor in 1979 by a group of workmen from Soochow in one of the light courts of the Metropolitan Museum of Art in New York, as a specimen of Chinese landscape art on permanent display.

In the Astor Court this Soochow garden, with material and workmanship leaving nothing to be desired, still prompts one to inquire, could a living organism like a garden be stored as any lifeless museum piece? Could a Chinese garden or part of it be transplanted to another continent to make the East-West rendezvous possible? It is one thing to reproduce a garden in brick and tile, but quite another to revive and kindle its life and spirit. In this connection, a higher criticism appeared in *Progressive Architecture* (Aug. 1981). It is worth quoting:

> *"Never the twain shall meet…The whole process was so undeniably sincere that it is heart-breaking how little the experience in Astor Court resembles walking in a genuine Chinese garden. The details are meticulous. One enters through a moon gate, and can follow a walkway that bends in and out, covered in a series of gray tile roofs. Latticed windows look out onto Chinese greenery. A spring bubbles around fantastically eroded Chinese rocks, and the walls are bisected by a perfectly detailed cornice as if they were exterior walls. But they are not. The court is indoors, built in an existing light well. The walls continue above the cornice and into a gabled skylight. The tile roofs butt against the walls and disappear. The bamboo beyond the lattice windows stand in shallow artificially lit niches. It is not question of slight translation. Essential to the long and subtle Chinese tradition is the presentation of each garden, indeed each element of each garden, as a piece of an infinitely extending landscape. Ingenious sleights of hand are attempted here, but they are resoundingly defeated by the pervasive walls and rectilinearity. Because here, under the skylight, what the Metropolitan presents is a charming quiet courtyard, hung and embedded with exquisite collected and*

reproduce fragments, just as Western museums always have. (N. M.)"

Going a step further than the Metropolitan, the Paris Center National D'Art et de Culture Georges Pompidou has made a model of this same Soochow garden at the scale of 1:50 to be exhibited at the Centre sometime in 1982. It thus will afford a bird's eye view and through electronic device may enable one to look at the arrangement of different parts of this garden. But is this enough? Remember what Goethe said: "Only when we walk around a building, move through it, can we share in its life". His remark would surely apply to a garden which has much more vitality and complexity than a mere building. To fully enjoy and appreciate the life of an original and not an imitative Chinese garden, one must participate in person, in strolling and pausing, in standing and sitting, in going into and out of every corner within the entire garden enclosure.

8. Hsi Yuan (西园, "West Garden"), not far from Liu Yuan, was in fact one time its companion garden, when in the sixteenth century the Hsu family, besides holding the East garden (now Liu Yuan), named this nearby garden, their detached property, Hsi Yuan, which was later donated to a monastery to become a temple garden. During the Tai-Ping war both temple and garden were destroyed, but soon rebuilt in 1892.

The central theme of the garden is the pond. In its midst stands a sizable arbor (fig. 32). The monks used the pond, as act of Buddhist benefaction, for setting free living fish that were caught by anglers elsewhere. The garden has long been isolated from the temple.

9. Ngou Yuan (耦园, "Twin Gardens") owes its beginning in the east part of the present ground to what first existed as She Yuan (涉园, "Stroll Garden") of Lu Chin (陆锦), a retired official in the early Manchu dynasty. It changed ownership several times afterwards and was finally bought by a retired mandarin Shen Ping-Cheng (沈秉成) about the year 1875. He entirely restored the old garden and besides, enlarged the whole property by incorporating the adjacent east compound where he built another smaller garden also. In it dominates a jewel-like library stack building, with lake rock adorning the courtyard. The coexistence of both east and west gardens prompted the owner to call the ensemble "Twin". In the days of the Republic the entire property came into the possession of a textile tycoon Liu. He held it until after the Liberation in 1950 and then donated it to the local government.

The twins are not balanced in size. The one on the east is much bigger and more elaborate, with its preeminently admirable yellow rock hill much of which had been built already in the early reign of Manchu dynasty. As a specimen full of natural charm, the hill is unrivaled by any other of its kind in the city.

10. I Pu (艺圃, 药圃, "Herb Garden") began as a herb garden in the early seventeenth century, built by Wen Chen-Meng (文震孟), great grandson of the famed

scholar and painter Wen Cheng-Ming (文征明,1470~1559). After the mid-seventeenth century its name was changed to I Pu ("Garden of the Craftsman").

This medium-sized garden is divided by a well-managed pool in the center of the landscape into a hill garden in the south and a group of buildings in the north. The hill, covered with exuberant planting and masterly arranged rock cliffs and paths, is a good specimen of design but suffers from wrong orientation, since its pictorial composition, looked from the north bank, stands against the light.

A small court, isolated by a wall, on the southwest of the pool, containing a streamlet and rockwork with buildings, is exquisite. On the pool's east bank stands the "Fry Arbor" (乳鱼亭, fig. 33), whose timber frame dates back to Ming dynasty.

11. He Yuan (鹤园, the "Stork"), one of the many small gardens in Soochow city, was built when the Manchu regime had but few more years to go, by a retired official Hung Lu-Ting (洪鹭汀).

A five-bay entrance hall, too ceremonial and outstanding perhaps, opens into the garden, and through series of zigzag verandas the visitor proceeds from one building to another. In the garden's center stands a pool, surrounded by rockery, architecture and planting.

Two features in the garden do much to relieve the monotony of three almost identical halls that dominate the scene. These are the "Trapezium pavilion" (梯形馆) adjoining the west wall and on its south, a small hill crowned by a hexagonal arbor,

32

with choice planting on all sides.

There are several more small gardens worth visiting. But one should not leave Soochow without seeing, on the slope of the Tiger Hill (虎丘), the terraced villa called Yung T'sui Shan Chuang.

12. Yung T'sui Shan Chuang (拥翠山庄), rare specimen of its kind so far known. It was built in 1884 by the Chu family and friends.

Within a difference of 11.5 meters in elevation, a strip of hilly ground is divided into four levels on which stand buildings, trees and rockeries. From the terraces, one commands unobstructed views far and wide, merging the garden with surrounding countryside and extending the

33

Fig. 32 A sizable arbor stands in mid-pond (Hsi Yuan)
Fig. 33 "Fry Arbor" (I Pu, Soochow)

图32 池水中央之硕大凉亭（西园）
图33 "乳鱼亭"（苏州艺圃）

landscape to the horizon, with the nearby pagoda inevitably in sight.

13. Kao Yi Yuan (高义园, "Garden of the Noble and Righteous") is another terraced garden located about 10 kilometers west of Soochow city, on the slope of a scenic hill known as Tien Ping Shan (天平山, "Skyhigh Hill"). The hill, well wooded with scarlet maple and orange and red chestnut oak in late autumn, once, as legend goes, afforded agreeable surroundings for the humble cottage of a young scholar and later a great statesman Fan Chung-Yan (范仲淹) of Northern Sung dynasty in the eleventh century. The property was enlarged by succeeding generation into a tiered villa, whose beauty is much enhanced by woods, springs, boulders and stone caves. The Manchu Emperor Chieh Lung visited the garden and to signify his admiration for the virtue and wisdom of Fan Chung-Yan, bestowed on the estate the title Kao Yi Yuan. The best season to enjoy the tour is the month of November.

14. Tsan Li Yuan (残粒园, "The Garden of Remnant Grain"). It was built in the last years of the Ching dynasty. Originally a small garden of a residence owned by a salt merchant, it has only one hundred square meters, with a dugout pond lined with lake rocks. Along the boundary wall, pavilions were built on hilltops. It is an exemplary small garden building.

• **Wukiang**

Tui Sze Yuan (退思园, "Retreat for Compunction") in Tung Li (同里), a village town, was owned by Jen Lan-Sheng (任兰

生), who after being dismissed from office through impeachment, retired to his home town to construct a house and garden in 1885, nursing feelings of repentance, on the site of a Sung dynasty villa. He asked the poet and painter Yuan Lung (袁龙) to design the garden, and to manifest his sense of humility, he laid stress on restraint, and as a result, the designer was prevented from displaying his full talent.

The central theme is the big hall with a pool in front. In the not too distant past, this garden used to be the gathering place of well known painters and poets, some of whom made a long stay.

• **Yangchow**

1. Ho Yuan (何园), poetically known as "Chi Shao Shan Chuang" (寄啸山庄, "Melody Conveying" Villa), was built on the site of an old garden by a retired Salt Gabelle administer Ho Chih-Fang (何芷舫) in the '80s of last century. It is the largest and best known private garden in Yangchow today, rivaling Soochow gardens in size and even in workmanship.

The central court, adorned with pool, arbor, rockery, adjoins an extensive two-story building in the background. A long, winding veranda shelters the inmate moving from building to building; group of houses close by once served as the owner's living quarter; and choice planting including some ancient trees in the garden is worthy of note.

2. Ke Yuan (个园), so called because the garden is dominated entirely by bamboo, and Ke (个) as a written character symbolizes graphically (竹, bamboo). It began as a garden in the early nineteenth century by a salt magnate Huang, whose penname was "Ke Yuan". To identify himself with his garden, he planted bamboo on the entire ground. Another unique feature is the "Rock Hill of Four Seasons" (四季假山). Four hills of piled stone, scattered over the garden ground, through sheer choice of the stone's color and species, disposal of shape and orientation, matched with appropriate planting, express the designer's skill in suggesting the feeling and phenomenon in each particular season of the mountain in nature. The four hills contain caves, light courts, steps, and from each top, one can look far into the countryside.

3. Ti Yuan (棣园) has existed since the early Manchu dynasty. It belonged successively to the families of Chen, Huang, Hung, and finally Pao, who named the garden Ti Yuan, in mid-nineteenth century. It was incorporated into the Hunan Guild House (湖南会馆)much later. The garden in its entirety was depicted in stone engraving in 1845 (fig. 34), being its copy in ink sketch.

4. Fu Chuang (凫庄, "Wild Duck" Villa), a tiny spot in the stream, was first built in 1921 by the Chen family, standing on a little island and connected with the shore by a bridge. The villa commands a fine panorama in every direction. Its entire layout today has just been completed anew after it fell in ruin in the '30s.

The above mentioned gardens in Yangchow are best known. Small and medium-sized gardens are numerous and only

Fig. 34 The garden was depicted in stone engraving (Ti Yuan, Yangchow)

图34　石刻园林图景（扬州棣园）

second to Soochow in quantity. Most are now in the process of repair and restoration.

Yangchow used to be the dwelling place of wealthy merchants, affluent officials and the noveau riche connected with salt trade. Their social activities required in the garden well-appointed halls for entertainment and open-air theatre for opera performance. Such special demands produced in Yangchow some features different from those gardens in Soochow and elsewhere.

• **Taichow**

Chiao Yuan (乔园). Its origin dates back to the sixteenth century when a high official in Ming dynasty Chen Ying-Fang (陈应芳) built on the site of She Yuan (涉园, "Stroll Garden"). In mid seventeenth century, the property passed to the Tien family. Some sixty years later it again changed hands. A Kao family possessed it, naming it Shan Feng Yuan (三峰园, "Garden

Glimpses of Gardens in Eastern China | 259

of Three Peaks Crowning the Rock Hill"). In about the '50s of last century, a rich salt administrator bought the garden, to become the last owner.

The garden's center of attraction is the five-bay hall with a U-shaped running stream on its south, bordered with rockery work. The stone-cave nearby was constructed in Ming dynasty. Since required amount of desirable rockery could not be acquired nearby, the expedient way to make an ornamental hill was either to use combination of lake rock and yellow stone or to cover a brick core with rockery veneer.

Directly north of the central hall is a separate court dominated by a hall flanked by a pavilion on each side. The garden is noted for old trees and a collection of fine flower plants.

• **Jukao**

Shui Hui Yuan (水绘园, "The Water-Graph Garden"). It was the Ming scholar Mao Pi-Chiang's (冒辟疆) residence of seclusion. With vast space and fine architecture, it once was a favorable haunt for many famous Ming scholars and later dwindled to a ruin. The present garden was reconstructed during the years of Chieh Lung by the Wangs. It has three suites separated by courts. The styles of this garden is close to that of the Yangzhou Gardens. The original name "The Water-Graph Garden" has been kept since its construction.

• **Nanking**

1. Chan Yuan (瞻园, "Look-up-to the Palace in the Moon Garden") started in the latter part of fourteenth century as the West Garden adjoining the princely residence of Hsu Ta (徐达), who helped the first Ming Emperor to win the empire. The most precious portion of the garden lies in the northeast, where stands a rockery hill with seven grottos below and a terrace above, sheltered by antique trees. The rockery all came reputedly from Tai Hu Lake near Soochow. Immediately in front of the hill is an irregularly shaped pond, connected with another smaller pond in the south by a canal, where a monumental hall is situated.

After the Ming dynasty was overthrown, the garden became part of the headquarters of the Manchu garrison commander. Emperor Chieh Lung, during a visit to the garden, expressed his pleasure by bestowing to it the title "Chan", and decreed it being duplicated in Peking, in the eastern part of the old Summer Palace. During the Tai-Ping war, it suffered damage but restoration began in 1867. Subsequent repairs were carried out in 1904, 1906 and 1942. The street entrance with fourcourt and rock hill was done in 1962.

2. Hsu Yuan (煦园, The Garden of "Solar Warmth") began at about the same time as Chan Yuan, being the princely garden of a relative of the first Ming Emperor. It is dominated by a rectangular pool, with a stone barge in midwater near its southern shore. Not far on the east on top of a rockery hill, stands an interlocking-square arbor, rarely found in garden architecture.

After the Manchu conquest, the garden and residential quarter on its east were occupied by the viceroy as the seat of his provincial government until the 1911 Revolution, interrupted only by the interval during the Tai-Ping war years (1853~1864), when its leader used the entire property as his royal palace. It suffered no damage and has since 1991 been occupied by one public body or another.

• **Shanghai**

1. Yu Yuan (豫园, Garden of "Ease and Happiness"), situated in the old city quarter, close to the Temple of the "City's Protecting God" (Cheng Huang Miao, 城隍庙), is the garden first built during 1560~1577 by a retired official Pan Yuen-Tuan (潘允端). On the south is Nei Yuan (内园, "Inner Garden"), began in 1709 as Tung Yuan (东园, "East Garden"). In 1760, Yu Yuan was also acquired by the Temple and to distinguish from its neighbor, was named West Garden (西园). Nei Yuan is only one eighth of the size of its companion. Yu Yuan has few rivals except in Soochow concerning extensive area and exquisitely planned buildings of remarkable workmanship. The yellow-stone hill in the northwestern corner is renowned as the creation of the mason Chang Nan-Yang (张南阳), a sixteenth-century rockwork expert. Close to the southern border, near a streamlet, stands a solitary rockery called "Yu Ling Lung" (玉玲珑, "Dexterously Carved Jade", fig. 35), a superb specimen of nature's abstract sculpture.

2. Nei Yuan, though small, is skillful in

Fig. 35 "Dextrously carved Jade" (Yu Yuan, Shanghai)

图35 "玉玲珑"（上海豫园）

planning, with architecture, hillock and streamlet agreeably disposed. Hardly any boundary wall is evident, since on the perimeter all walls were used as exterior of buildings.

- **Nanziang**

I Yuan (猗园, "Exuberant and Beautiful" Garden), originally a garden of the Min family in the sixteenth century, later passed into the hands of the Li family and was named I Yuan. In 1748, a new owner, Yeh, after having acquired the property and done repairs, renamed it "Ku (ancient) I Yuan" (古猗园). The garden was donated to a temple in 1788. It later passed into private ownership, and in 1808 was acquired and repaired by yet another family. After the Tai-Ping war restoration was carried out in 1868. The year 1921 saw the garden in complete dilapidation, but it soon was revived in the '30s.

In 1980, the garden's layout began to be entirely overhauled. A new study was being carried out, step by step, on building, planting, promenade, and hill and water management, for the purpose of accommodating, with an up-to-date landscape recreation ground ready, the large crowd pouring in from Shanghai on holiday.

- **Wusih**

Chi Ch'ang Yuan (寄畅园, "Delight Conveying" Garden) was first built in the early sixteenth century by an official Ch'in Chin(秦金), who chose the hill Hui Shan (惠山) as the background of his garden founded on the site of a temple. His descendants inherited the property and his great grandson named the garden "Chi Ch'ang". The Manchu Emperors Kang Hsi and Chieh Lung both visited the garden during their southern tours. The garden is noted for its many springs under the pool and on ground surface. The spring water was exploited to the full by the garden's designer to murmur in the rockwork or to gush forth under a terrace. Springs in the neighborhood, in the Ming dynasty, also nourished no less than eight gardens which have long since disappeared.

To the south of Wusih city, near the shore of Lake Tai Hu, there are several gardens mostly owned in the '30s by wealthy industrialists.

- **Changchow**

Chin Yuan (近园, "Somewhat a Garden") was first built in the early Manchu dynasty by a Scholar-Mandarin Yang Chao-Lu (杨兆鲁) who modestly expressed satisfaction even if his garden was far from being considered perfect. A rock hill with grotto stands in the pool behind which one finds the main hall. Other buildings on the periphery of the garden are all connected by covered corridors.

During the 1670 famed scholars and painters were constant guests in this garden. It was sold to the Lius in about 1870 and some years later was bought by the Tun family, its last owner.

- **Quinsan**

Pan Chien Yuan (半茧园, "Half-Cocoon" Garden) was once the garden of

Yeh family first built in 1546. Succeeding generations enlarged it, finally naming it Chien (Cocoon). In the Manchu dynasty the garden was divided up and one of Yeh's descendants occupied half of it, calling it "Half Cocoon". Later the garden was purchased by the Lu family. In mid-eighteenth century the property was turned over to a temple, enriched by several new garden buildings and in addition, walled in all around the boundary. Alterations and repairs were done in 1801 and again in 1823. During the interval a piece of rockery (寒翠石, "Cool Green Rock") was moved to the garden's hill top from a nearby village. The rock, a specimen of abstract sculpture, dated its appearance in the eleventh century A.D. (fig. 36).

Fig. 36 "Cool Green Rock" (Pan Chien Yuan, Quinsan)

图36 "寒翠石"（昆山半茧园）

• **Sungkiang**

Tsui Pei Ch'ih (醉白池, Garden of the "Intoxicated Poet"). Sungkiang was for a long time the prefectural city to which Shanghai was subordinate. In the Manchu dynasty, the city abounded with gardens, which, however disappeared one after another. Now Tsui Pei Ch'ih remains as the only attraction. This garden was started by a painter named Ku during the reign of the Manchu Emperor K'ang Hsi (1662~1722). In the early nineteenth century, it belonged to a founding asylum.

• **Chingpu**

Chu Shui Yuan (曲水园, "Winding Stream") was first built by the Han family in 1745 as a temple garden. At the beginning of nineteenth century, it started to

37

assume the title "Chu Shui", famous for its twenty-four sceneries. The garden was destroyed in 1860, but restoration took place in late nineteenth century.

• **Kating**

Chiu Hsia Pu (秋霞圃, "Autumn Sunset") was started in the early sixteenth century by high government official Kung Hung (龚弘). In the mid seventeenth century, it became the property of Wang, a nouveau riche, who did much to make the garden elegant and attractive. The property was turned over to the City Temple in 1726. Some forty years later, this garden was combined with the East Garden of Shen family. During the Tai-Ping uprising

Fig. 37~38 I Yuan, Taichang
(Photographed in 1930's)

图37~38 太仓亦园（摄于20世纪30年代）

38

much damage was done, and badly needed repair work was carried out in 1886.

• **Chiangyin**

Shih Yuan (适园, "Easy and Comfortable" Garden) belonged to a Manchu dynasty Han-Lin Scholar Chen who built it between the years 1854~1864. Also a poet and painter, he designed half of the site to a yellow stone hill. South of the hill is the water pond. On the shore and facing the hill stands the main hall, connected by a covered passage with the owner's studio to the east. Another hall is behind the hill on the north. This medium-sized garden has a rich variety of architecture and planting seldom found in many other gardens. In the 1970s extensive restoration and repair unavoidably sacrificed some of the garden's original charm.

• **Taichang**

Taichang in the '30s had two or three dilapidated gardens, but a portion of I Yuan (亦园, "Also a Garden") of Chiang family still stood in good condition with exquisite architecture (fig. 37~38) even though the property had been requisitioned as part of a hospital. All gardens, though, disappeared during World War II. Today the local authorities are planning to restore some of them.

- **Changshu**

1. Yen Yuan (燕园, Yen Ku, 燕谷, "Swallow's Gorge") was built by Chiang Yuan-Shu (蒋元枢), a mandarin retiring from his post as chief administrator of Formosa in 1780. Fifty years later, it passed to his clansman Chiang Yin-Pei (蒋因培), who commissioned the mason Ke Yu-Liang (戈裕良) of rockery fame, to construct the water-side hill known as Yen Ku, to remind himself of being relieved from official duty and enjoying home–coming to Changshu, like a swallow returning to its nest. The garden was sold to Kuei family in 1847. Near the end of nineteenth century, Chiang family bought it back again, but soon Chang Hung (张鸿), a Hanlin Scholar, took possession of it as the last owner, and did extensive rehabilitation work. Choice planting and tasteful craftship rank this garden among one of the best in Southeast China.

2. Hsu Kuo Chu (虚廓居, "Garden for the Vacant Mind") was built on the ruined site of a Ming dynasty garden in about 1894 by Tseng Chih-Chuan (曾之撰), a literary man and retired official. Waterscape dominates the scene, just the inviting element in summer time that gives rise to tea-drinking pavilions.

3. The Tings' Garden (丁氏园). It was built by Ting Kui-Ching (丁葵忱), a selected candidate in the provincial Imperial Examinations of late Qing dynasty. It is a small garden nestling close to the house, with only a palm-size of ground. It has a mini pond of four or five square meters, joined by a stream that flows along its eastside. Pieces of lake rocks have been planted in the garden which is joined to the house by winding corridor. It is one of the few surviving private small gardens.

- **Chekiang**

Chekiang province is equally famed for its traditional gardens, which, though, are less numerous than those in Kiangsu. Hangchow, the well known resort city, has lost most if not all of its fine villas and gardens except Kuo Chuang (郭庄) in the process of urban modernization.

- **Kashing**

Kashing is left with Yen Yu Lou (烟雨楼, "Misty Tower") in the lake. It dates back to the tenth century. The tower commands fine views and is used as a teahouse. Outside the city to the north is Lo Fan Ting (落帆亭, "Under Bare Pole" Pavilion), a small garden near an ancient canal lock, dating its origin to Sung dynasty. Nanziang, until the end of the '30s, as a small but wealthy town, was celebrated for its gardens like Yee Yuan (宜园) of Pang family, Shih Yuan (适园) of Chang family, and Hsiao Lien Chuang (小莲庄) of Liu family, which later being best known for its owner's book collection, housed in a library in the garden.

- **Haiyan**

Chi Yuan (绮园, "Beautiful Garden") was built on the site of earlier house and garden in early Manchu dynasty. In the 19th century, it belonged to an official and scholar Huang, who after repair and restoration named the garden Chun Yi (拙宜园,

"Garden Good for the Stupid"), only to be destroyed in the Tai-Ping wars. His son-in-law Feng inherited the empty lot with only an ancient tree on the ground, and began hill and water planning, naming it Chi Yuan. But lack of fund and scholarly taste lent the garden no charm. Only two ingenious features distinguish this garden from the ordinary: a stone slab between tops of two perpendicular cliffs as a flying bridge and a causeway in midwater to imitate the West Lake in Hangchow. Just now the garden is being under repair.

◆ Pinghu

Mo Garden (莫氏园). The Mos, natives of Fuchien Province, in the early Manchu dynasty made a fortune in timber trade. One member of the family, Mo Fang-Mei (莫放梅, 1827~1889) settled in Shanghai, but still maintained connection with Pinghu, and after buying an empty lot, built his house and garden there to strengthen the family ties.

The garden consists of two small oblong yards serving as light courts. In the courtyards one finds plating, a small basin and some outdoor furniture. The courts adjoin a maze of buildings, some of which are two storied.

plant life thrive in one garden just as well as in any other. Rockery hill, lily pond and water pavilion are common everywhere. Most gardens, if not owned by, at least were frequented by scholars, poets, and painters. It is also worthy of note, as pointed out before, that gardens in Soochow have never been surpassed in beauty and in number by any other district.

From the gardens in some dozens of localities cited above, we have a collection of the most sublime specimens of Chinese landscape art. All situated in the temperate zone, garden architecture is endowed by mild climate with light construction and delicate detail, and the same species of

附录
Appendix

附录一
Appendix I

园林植物系谱(苏州地区)
Classified List of Garden Plants (Soochow Region)

A
观花类
Blossom Type

常青类 Evergreen

1	山 茶	*Camellia japonica* L.
2	瓶 兰	*Diospyros armata* Hemsl.
3	栀子花	*Gardenia jasminoides* Ellis
4	金丝桃	*Hypericum chinense* L.
5	探 春	*Jasminum floridum* Bunge
6	云南迎春	*Jasminum mesnyi* Hance
7	广玉兰	*Magnolia grandiflora* L.
8	含 笑	*Michelia figo* (Lour.) Spreng
9	夹竹桃	*Nerium indicum* Mill.
10	桂 花	*Osmanthus fragrans* Lour.
11	丹 桂	*O.fragrans var.aurantiacus* Makino
12	花石榴	*Punica granatum var.nana* Pers.
13	杜 鹃	*Rhododendron simsii* Planch
14	月 季	*Rosa chinensis* Jacq.
15	六月雪	*Serissa serissoides* (DC.) Druce

落叶类 Deciduous

16	锦鸡儿	*Caragana sinica* (Buc'hoz) Rehd.
17	紫 荆	*Cercis chinensis* Bunge
18	木 瓜	*Chaenomeles sinensis* (Thouin) Koehne
19	贴梗海棠	*Chaenomeles speciosa* (Sweet) Nakai
20	蜡 梅	*Chimonanthus praecox* Link.
21	连 翘	*Forsythia suspensa* Vahl.
22	金钟花	*Forsythia viridissima* Lindl.
23	木芙蓉	*Hibiscus mutabilis* L.

	24	扶　桑	*Hibiscus rosa-sinensis* L.
	25	木　槿	*Hibiscus syriacus* L.
	26	八仙花	*Hydrangea macrophylla* (Thunb.) Seringe
	27	迎　春	*Jasminum nudiflorum* Lindl.
	28	棣　棠	*Kerria japonica* (L.) DC.
	29	重瓣棣棠	*Kerria japonica* f. *pleniflora* (Witte) Rehd.
	30	紫　薇	*Lagerstroemia indica* L.
	31	银　薇	*Lagerstroemia indica* f. *alba* Nichols
	32	白玉兰	*Magnolia denudata* Desr.
	33	辛夷(木笔)	*Magnolia liliflora* Desr.
	34	二乔玉兰	*M.soulangeana* Soul.
	35	垂丝海棠	*Malus halliana* Koehne
	36	西府海棠	*Malus micromalus* Makino
	37	海　棠	*Malus spectabilis* (Ait.) Borkh.
	38	牡　丹	*Paeonia suffruticosa* Andrews
	39	杏　花	*Prunus armeniaca* L.
	40	郁　李	*Prunus japonica* Thunb.
	41	梅　花	*Prunus mume* Sieb.et Zucc.
	42	绿萼梅	*Prunus mume* var. *viridicalyx* Makino
	43	桃　花	*Prunus persica* (L.) Batsch
	44	碧　桃	*Prunus persica* f. *duplex* (West.) Rehd.
	45	白花碧桃	*Prunus persica* f. *alba-plena* Schneid.
	46	红花碧桃	*Prunus persica* f. *camelliaeflora* Dipp.
	47	李	*Prunus salicina* Lindl.
	48	日本晚樱	*Prunus serrulata* var. *lannesiana* (Carr.) Rehd.
	49	榆叶梅	*Prunus triloba* Lindl.
	50	日本樱花	*Prunus yedoensis* Matsum.
	51	珍珠梅	*Sorbaria kirilowii* Maxim
	52	麻叶绣球	*Spiraea cantoniensis* Lour.
	53	欧洲丁香	*Syringa vulgaris* L.
	54	绣球花	*Viburnum macrocephalum* Fort.
	55	锦带花	*Weigela florida* (Bge) A.DC.

B 果类 Fruit Type	常青类 Evergreen		
	1	代代橘	*Citrus aurantium* var. *amara* Engl.
	2	佛　手	*Citrus medica* var. *sarcodactylis* (Noot.) Swingle

	3	橘	*Citrus reticulata* Blanco
	4	枇杷	*Eriobotrya japonica* (Thunb.) Lindl.
	5	枸骨	*Ilex cornuta* Lindl.
	6	南天竹	*Nandina domestica* Thunb.
	7	珊瑚树	*Viburnum odoratissimum* Ker-Gawl.

落叶类 Deciduous

8	柿	*Diospyros kaki* Thunb.
9	无花果	*Ficus carica* L.
10	枸杞	*Lycium chinense* Mill.
11	花红	*Malus asiatica* Makai
12	石榴	*Punica granatum* L.
13	枣	*Zizyphus jujuba* Mill.

C
观叶类
Foliage Type

常青类 Evergreen

1	青木	*Aucuba japonica* Thunb.
2	黄杨	*Buxus sinica* Rehd.et Wils.
3	八角金盘	*Fatsia japonica* Decne.et Planch.
4	女贞	*Ligustrum lucidum* Ait.
5	海桐	*Pittosporum tobira* Ait.
6	棕榈	*Trachycarpus fortunei* H. Wendl.
7	丝兰	*Yucca filamentosa* L.
8	凤尾兰	*Yucca gloriosa* L.

落叶类 Deciduous

9	鸡爪槭	*Acer palmatum* Thunb.
10	红枫	*Acer palmatum var. atropurpureum* (Vanh.) Schwer.
11	蓑衣槭	*Acer palmatum var. thunbergii* pax.
12	山麻杆	*Alchornea davidii* Franch.
13	枫香	*Liquidambar formosana* Hance
14	红叶李	*Prunus cerasifera atropurpurea* (Jaeg.) Rehd
15	垂柳	*Salix babylonica* L.
16	乌桕	*Sapium sebiferum* (L.) Roxb.
17	柽柳	*Tamarix chinensis* Lour.

D		常青类 Evergreen		
遮荫类		1	香 樟	*Cinnamomum camphora* (L.) Presl
For Shade		2	柳 杉	*Cryptomeria fortunei* Hooibrenk ex Ott et Dietr.
		3	白皮松	*Pinus bungeana* Zucc.
		4	马尾松	*Pinus massoniana* Lamb.
		5	黑 松	*Pinus thunbergii* Parl.
		6	罗汉松	*Podocarpus macrophyllus* (Thunb.) D. Don.
		7	圆 柏	*Sabina chinensis* (L.) Antoine

落叶类 Deciduous

		8	臭 椿	*Ailanthus altissima* Swingle
		9	合 欢	*Albizzia julibrissin* Durazz.
		10	糙叶树	*Aphananthe aspera* (Thunb.) Planch.
		11	梓 树	*Catalpa ovata* D. Don
		12	朴 树	*Celtis sinensis* Pers.
		13	梧 桐	*Firmiana simplex* (L.) W.F.Wight
		14	银 杏	*Ginkgo biloba* L.
		15	皂 荚	*Gleditsia sinensis* Lam.
		16	楝 树	*Melia azedarach* L.
		17	黄连木	*Pistacia chinensis* Bge.
		18	枫 杨	*Pterocarya stenoptera* DC.
		19	槐 树	*Sophora japonica* L.
		20	榔 榆	*Ulmus parvifolia* Jacq.
		21	白 榆	*Ulmus pumila* L.
		22	榉 树	*Zelkova schneideriana* Hand-Mazz.

E		常青类 Evergreen		
攀缘植物		1	三角花	*Bougainvillea glabra* Choisy
Climbing Plants		2	薜 荔	*Ficus pumila* L.
		3	常春藤	*Hedera helix* L.
		4	金银花 (忍冬)	*Lonicera japonica* Thunb.
		5	黄木香	*Rosa banksiae* f. *lutescens* Voss.
		6	白木香	*Rosa banksiae* var. *normalis* Reg.
		7	蔷 薇	*Rosa multiflora* Thunb.

8	铺地柏	*Sabina procumbens* (Endl.) Iwata et Kusaka
9	络 石	*Trachelospermum jasminoides* (Lindl.) Lem.

落叶类 Deciduous

10	凌 霄	*Campsis grandiflora* (Thunb.) Loisel.
11	美国凌霄	*Campsis radicans* (L.) Seem.
12	爬山虎	*Parthenocissus tricuspidata* Planch.
13	葡 萄	*Vitis vinifera* L.
14	紫 藤	*Wisteria sinensis* Sweet
15	白花紫藤	*W.sinensis var. alba* Lindl.

F
竹类
Bamboo Family

1	观音竹	*Bambusa multiplex (Lour.)* Raeuschel
2	方 竹	*Chimonobambusa quadrangularis* (Fenzi) Makino
3	箬 竹	*Indocalamus tessellatus* (Munro) Keng f.
4	斑 竹	*Phyllostachys bambusoides f. tanakae* Makino ex Tsuboi
5	紫 竹	*Phyllostachys nigra* (Lodd.) Munro
6	净 竹	*Phyllost nuda* McClure
7	罗汉竹	*Phyllost pubescens var. hererocycla* (Carr.) H.de Lehaie
8	碧玉间黄金竹	*Phyllost viridis* (Young) McClure cv. HOUZEAU
9	慈 竹	*Sinocalamus affinis* (Rendle) McClure

G
草本植物，水生植物
Herbaceous Plants, Aquatic Plants

1	秋 葵	*Abelmoschus esculentus* (L.) Moench
2	蜀 葵	*Althaea rosea* Cav.
3	雁来红	*Amaranthus tricolor* L.
4	秋海棠	*Begonia*
5	鸡冠花	*Celosia cristata* L.
6	菊 花	*Chrysanthemum morifolium* Ramat.
7	君子兰	*Clivia miniata* Regel
8	鸭跖草	*Commelina communis* L.

9	香雪兰	*Freesia refracta* Klatt
10	倒挂金钟	*Fuchsia hybrida* Voss.
11	萱草	*Hemerocallis fulva* L.
12	朱顶兰	*Hippeastrum vittatum* Herb.
13	玉簪	*Hosta plantaginea* Aschers
14	紫萼	*Hosta ventricosa* Stearn
15	凤仙花	*Impatiens balsamina* L.
16	鸢尾	*Iris tectorum* Maxim.
17	山麦冬	*Liriope spicata* (Thunb.) Lour.
18	紫茉莉	*Mirabilis jalapa* L.
19	芭蕉	*Musa basjoo* Sieb.et Zucc.
20	荷花	*Nelumbo nucifera* Gaertn.
21	萍蓬草	*Nuphar pumilum* (Timm.) DC.
22	睡莲	*Nymphaea tetragona* Georgi
23	书带草	*Ophiopogon japonicus* Ker-Gawl.
24	诸葛菜	*Orychophragmus violaceus* (L.) O.E.Schulz
25	芍药	*Paeonia lactiflora* Pall.
26	芦苇	*Phragmites communis* Trin.
27	虎耳草	*Saxifraga stolonifera* (L.) Meerb

袁以苇整理（南京中山植物园）

Compiled by Yuan Ih-Wei

Nanjing Botanical Gardern Mem. Sun Yat-sen

附录二 名称韦氏/邮政式拼写及拼音
Appendix II Names in WADE / POSTAL SPELLING and PINYIN

Name	WADE / POSTAL SPELLING	PINYIN
童 寯	Tung Chuin	Tong Jun
刘敦桢	Liu Tun-Tseng	Liu Dunzhen
顾辟疆	Ku Pi-Chiang	Gu Pijiang
王 维	Wang Wei	Wang Wei
杜 绾	Tu Wan	Du Wan
张南垣	Chang Nan-Yuan	Zhang Nanyuan
戈裕良	Ke Yu-Liang	Ge Yuliang
徽 宗	Hui Tsung	Hui Zong
乾 隆	Chieh Lung	Qian Long
桀	Chieh	Jie
文 王	Wen Wang	Wen Wang
始皇帝	Shih Huang-Ti	Shi Huangdi
武 帝	Wu Ti	Wu Di
梁孝王	Liang Hsiao-Wang	Liang Xiao Wang
袁广汉	Yuan Kuang-Han	Yuan Guanghan
董仲舒	Tung Chung- Shu	Dong Zhongshu
石 崇	Shih Chung	Shi Chong
王羲之	Wang Hsi-Chih	Wang Xizhi
白居易	Po Chu-Yi	Bai Juyi
郭忠恕	Guo Chung-Shu	Guo Zhongshu
郭士元	Guo Shi-Yuan	Guo Shiyuan
李格非	Li Ke-Fei	Li Gefei
朱 勔	Chu Mien	Zhu Mian
朱伯原	Chu Peh-Yuan	Zhu Baiyuan
计 成	Chi Ch'eng	Ji Cheng
朱舜水	Chu Shun-Shui	Zhu Shunshui

Name	WADE / POSTAL SPELLING	PINYIN
康 熙	K'ang Hsi	Kang Xi
袁 枚	Yuan Mei	Yuan Mei
惟 则	Wei Tse	Wei Ze
倪 瓒	Ni Tsan	Ni Zan
杜东原	Tu Tung-Yuan	Du Dongyuan
申时行	Shen Shih-Hsin	Shen Shixing
顾文彬	Koo Wen-Pin	Gu Wenbin
苏舜钦	Shu Shun-Chin	Su Shunqin
文 瑛	Wen Ying	Wen Ying
史正志	Shih Cheng-Chih	Shi Zhengzhi
宋宗元	Sung Tsung-Yuan	Song Zongyuan
瞿远村	Chu Yuan-T'sun	Qu Yuancun
李鸿裔	Li Hung-I	Li Hongyi
达 桂	Ta Kuei	Da Gui
张金坡	Chang Chin-Po	Zhang Jinpo
何亚农	Ho Ya-Nung	He Yanong
陆 锦	Lu Chin	Lu Jin
沈秉成	Shen Ping-Cheng	Shen Bingcheng
文震孟	Wen Chen-Meng	Wen Zhenmeng
文征明	Wen Cheng-Ming	Wen Zhengming
洪鹭汀	Hung Lu-Ting	Hong Luting
范仲淹	Fan Chung-Yan	Fan Zhongyan
任兰生	Jen Lan-Sheng	Ren Lansheng
袁 龙	Yuan Lung	Yuan Long
何芷舫	Ho Chih-Fang	He Zhifang
陈应芳	Chen Ying-Fang	Chen Yingfang
徐 达	Hsu Ta	Xu Da
潘允端	Pan Yuen-Tuan	Pan Yunduan
张南阳	Chang Nan-Yang	Zhang Nanyang
秦 金	Ch'in Chin	Qin Jin
杨兆鲁	Yang Chao-Lu	Yang Zhaolu
龚 弘	Kung Hung	Gong Hong
蒋元枢	Chiang Yuan-Shu	Jiang Yuanshu
蒋因培	Chiang Yin-Pei	Jiang Yinpei
张 鸿	Chang Hung	Zhang Hong
曾之撰	Tseng Chih-Chuan	Zeng Zhizhuan

Name	WADE / POSTAL SPELLING	PINYIN
丁葵忱	Ting Kui-Ching	Ding Kuichen
冒辟疆	Mao Pi-Chiang	Mao Pijiang
莫放梅	Mo Fang-Mei	Mo Fangmei
袁以苇	Yuan Ih-Wei	Yuan Yiwei
晏隆余	Yan Lung-Yu	Yan Longyu

苏 州	Soochow	Suzhou
扬 州	Yangchow	Yangzhou
南 京	Nanking	Nanjing
常 州	Changchow	Changzhou
泰 州	Taichow	Taizhou
如 皋	Jukao	Rugao
上 海	Shanghai	Shanghai
南 翔	Nanziang	Nanxiang
无 锡	Wusih	Wuxi
杭 州	Hangchow	Hangzhou
昆 山	Quinsan	Kunshan
松 江	Sungkiang	Songjiang
青 浦	Chingpu	Qingpu
嘉 定	Kating	Jiading
江 阴	Chiangyin	Jiangyin
太 仓	Taichang	Taicang
吴 江	Wukiang	Wujiang
常 熟	Changshu	Changshu
嘉 兴	Kashing	Jiaxing
南 浔	Nanzing	Nanxun
平 湖	Pinghu	Pinghu
海 盐	Haiyan	Haiyan
太 湖	Tai Hu	Tai Hu
洛 阳	Loyang	Luoyang
长 安	Ch'angan	Chang'an
西 安	Sian	Xi'an
河 南	Honan	Henan
河 阳	Hoyang	Heyang
开 封	Kaifeng	Kaifeng

Name	WADE / POSTAL SPELLING	PINYIN
湖州	Huchow	Huzhou
吴兴	Wuhsing	Wuxing
江苏	Kiangsu	Jiangsu
浙江	Chekiang	Zhejiang
天目山	Tien Mu Shan	Tianmu Shan
景德路	Ching Teh Road	Jingde Lu
同里	Tung Li	Tongli
城隍庙	Cheng Huang Miao	Chenghuang Miao
惠山	Hui Shan	Hui Shan
辋川	Wang Ch'uan	Wangchuan
环秀山庄	Huan Hsin Shan Chuang	Huanxiu Shanzhuang
艮岳	Ken Yao	Gen Yue
北海	Peh Hai	Beihai
上林	Shang Lin	Shang Lin
小金山	Hsiao Chin Shan	Xiaojin Shan
平山堂	Ping Shan Tang	Pingshan Tang
煦园	Hsu Yuan	Xu Yuan
瞻园	Chan Yuan	Zhan Yuan
随园	Sui Yuan	Sui Yuan
拙政园	Chueh Cheng Yuan	Zhuozheng Yuan
复园	Fu Yuan	Fu Yuan
狮子林	Shih Tzu Lin	Shizi Lin
留园	Liu Yuan	Liu Yuan
汪义庄	Wang Yi Chuang	Wangyi Zhuang
怡园	Yi Yuan	Yi Yuan
沧浪亭	T'sang Lang Ting	Canglang Ting
网师园	Wang Shih Yuan	Wangshi Yuan
西园	Hsi Yuan	Xi Yuan
耦园	Ngou Yuan	Ou Yuan
涉园	She Yuan	She Yuan
艺圃	I Pu	Yi Pu
鹤园	He Yuan	He Yuan
拥翠山庄	Yung T'sui Shan Chuang	Yongcui Shanzhuang
高义园	Kao Yi Yuan	Gaoyi Yuan
天平山	Tien Ping Shan	Tianping Shan
残粒园	Tsan Li Yuan	Canli Yuan

Name	WADE / POSTAL SPELLING	PINYIN
退思园	Tui Sze Yuan	Tuisi Yuan
何　园	Ho Yuan	He Yuan
寄啸山庄	Chi Shao Shan Chuang	Jixiao Shanzhuang
个　园	Ke Yuan	Ge Yuan
棣　园	Ti Yuan	Di Yuan
凫　庄	Fu Chuang	Fu Zhuang
乔　园	Chiao Yuan	Qiao Yuan
三峰园	Shan Feng Yuan	Sanfeng Yuan
豫　园	Yu Yuan	Yu Yuan
内　园	Nei Yuan	Nei Yuan
东　园	Tung Yuan	Dong Yuan
猗　园	I Yuan	Yi Yuan
古漪园	Ku I Yuan	Guyi Yuan
寄畅园	Chi Ch'ang Yuan	Jichang Yuan
近　园	Chin Yuan	Jin Yuan
半茧园	Pan Chien Yuan	Banjian Yuan
醉白池	Tsui Pei Ch'ih	Zuibai Chi
曲水园	Chu Shui Yuan	Qushui Yuan
秋霞圃	Chiu Hsia Pu	Qiuxia Pu
适　园	Shih Yuan	Shi Yuan
亦　园	I Yuan	Yi Yuan
燕　园	Yen Yuan	Yan Yuan
虚廓居	Hsu Kuo Chu	Xukuo Ju
丁氏园	Ting Shih Yuan	Dingshi Yuan
郭　庄	Kuo Chuang	Guo Zhuang
烟雨楼	Yen Yu Lou	Yanyu Lou
落帆亭	Lo Fan Ting	Luofan Ting
宜　园	Yee Yuan	Yi Yuan
小莲庄	Hsiao Lien Chuang	Xiaolian Zhuang
绮　园	Chi Yuan	Qi Yuan
拙宜园	Chun Yi Yuan	Zhuoyi Yuan
水绘园	Shui Hui Yuan	Shuihui Yuan

编者注：书中第155、175、177、185、193、195、197、203—216、280页的图由摄影师青简提供，其余彩图由摄影师郑可俊提供，英文部分的黑白图由译者童明提供。

图书在版编目（CIP）数据

东南园墅 / 童寯著；童明译 . -- 长沙：湖南美术出版社，2018.10
ISBN 978-7-5356-8412-7

Ⅰ.①东⋯ Ⅱ.①童⋯ ②童⋯ Ⅲ.①园林艺术−研究 Ⅳ.①TU986.1

中国版本图书馆CIP数据核字（2018）第182805号

东南园墅
DONGNAN YUANSHU

童寯 著　　童明 译

出 版 人	黄啸
出 品 人	陈垦
出 品 方	中南出版传媒集团股份有限公司
	上海浦睿文化传播有限公司（上海市巨鹿路417号705室 200020）
责任编辑	张抱朴
责任印刷	王磊
出版发行	湖南美术出版社（长沙市雨花区东二环一段622号）
网　　址	www.arts-press.com
经　　销	湖南省新华书店
印　　刷	深圳市福圣印刷有限公司

开本：889mm×1194mm　1/16　　印张：18.25　　字数：110千字
版次：2018年10月第1版　　　　　印次：2023年7月第8次印刷
书号：ISBN 978-7-5356-8412-7
　　　　　　　　　　　　　　　　定价：96.00元

版权专有，未经本社许可，不得翻印。
如有倒装、破损、少页等印装质量问题，请与印刷厂联系调换。
联系电话：8621-60455819

出 品 人：陈　垦
策 划 人：蔡　蕾
监　　制：余　西　刘　佳
出版统筹：戴　涛
编　　辑：林晶晶
摄　　影：郑可俊　青　简
装帧设计：杨林青工作室

投稿邮箱：insightbook@126.com
新浪微博：@浦睿文化